D1566992

The Cambridge Astronomy Guide

A practical introduction to astronomy

William Liller and Ben Mayer

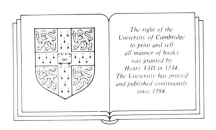

Cambridge University Press

Cambridge
London   New York   New Rochelle
Melbourne   Sydney

Published by the Press Syndicate of the University of Cambridge
The Pitt Building, Trumpington Street, Cambridge CB2 1RP
32 East 57th Street, New York, NY 10022, USA
10 Stamford Road, Oakleigh, Melbourne 3166, Australia

© Cambridge University Press 1985

First published 1985

Printed in Great Britain

British Library cataloguing in publication data

Liller, William
   Cambridge astronomy guide: the amateur and
professional viewpoints.
   1. Astronomy – Observers' manuals.
   I. Title II. Mayer, Ben
522 QB64

Library of Congress cataloguing in publication data
Liller, William
   Cambridge astronomy guide.
   Bibliography: p.
   Includes index.
   1. Astronomy – Amateurs' manuals. I. Mayer,
Ben II. Title.
QB63. M444 1985   523   85–16582
ISBN 0 521 25778 6

Acknowledgements

We should like to thank all those who have contributed photographs to this book.
Figs 1.1, 2.1, 2.5, 2.9, 2.10, 3.2, 4.1, 5.2, 7.1, 8.2, 8.10, 9.1, 9.2, 10.1, 10.4, 11.1, 13.6, 14.1, 14.3, 15.1, 16.1, 16.6, 16.7, 17.1, 19.1, 19.8, 19.9, 21.1, 21.2, 21.3, 21.4, 21.5, 21.6, 21.7, 21.8, 21.9, 21.10, 21.11, 21.12, 21.13, 21.14 © Ben Mayer.
Fig. 2.6 courtesy of the Smithsonian Astrophysical Observatory.
Fig. 2.8 © 1981 Doug Johnson, courtesy of the US National Radio Astronomy Observatory, operated by Associated Universities, Inc. under contract with the National Science Foundation.
Fig. 5.1 © Manuel Lopez Alvarez.
Figs 8.1, 10.7 © Dennis Milon.
Fig. 8.3 © Sherman Schultz.
Figs 8.4, 20.3 © Hilary Liller.
Fig. 10.8 © Marcelo Bass and Oscar Saa; University of Michigan Schmidt telescope at the Cerro Tololo Interamerican Observatory, operated by AURA, Inc. under contract with the National Science Foundation.
Fig. 12.1 courtesy of NASA.
Fig. 16.2 courtesy of the Canada–France–Hawaii Telescope Corporation.
Fig. 16.3 courtesy of Arecibo Observatory.
Fig. 16.4 courtesy of Dr R. M. West.
Fig. 16.5 © ESO.
Figs 17.2, 19.1 © James Wray, McDonald Observatory.
Fig. 19.11 © Lucille Mayer.
Fig. 20.1 courtesy of Optec, Inc.
Fig. 20.2 courtesy of Stano Components.

STARFRAMES and PROBLICOM are trademarks of Ben Mayer.

Contents

Acknowledgements		2
Preface to the odd chapters		4
Preface to the even chapters		5
1	Transcendental astronomy	6
2	First things first: fundamental thoughts and concepts	10
3	The patient eye – the camera	22
4	More about cameras – telescopes too	26
5	Catching starlight	34
6	About film	40
7	A picture a night, a film a month	46
8	Where to look, what to shoot	52
9	From the wastebasket to the Smithsonian	62
10	Understanding your camera subjects	68
11	Buy half a telescope	78
12	The Earth in motion	84
13	Two down, three across	90
14	Scientific fallout: the contributions of the amateur	98
15	The blinking astronomer	108
16	A night in a mountaintop observatory	116
17	Go ahead, get that 'scope . . .	126
18	Choosing and using a telescope	132
19	Do your own thing . . .	138
20	Closed-circuit TV and other aids	146
21	Miracle objects	150
22	Space for amateurs	166
Appendix		170
Resources		171
Index		172

PREFACE TO THE ODD CHAPTERS

It is not surprising that even casual stargazers are soon captivated by the enchantment of the night sky. As soon as one probes further into the treasures of the Galaxy, the rapture of deep space casts its spell and one is transported to the very edge of time beyond all known dimensions. That is when astronomy changes from a pursuit into a love affair, from a science into a passion.

In terms of what there is to know about the inscrutable mysteries of the unfathomable beyond the stars, the most learned professor of astronomy and the humblest beginning amateur know equally little. In fact the more we learn, the sooner we come to realize the limitations of our own faculties. This can lead us to an inner awareness of a superior order to which we all belong. It can guide the way toward an understanding for a mystical spirit which guides all things.

Since the beginning of time, through the ages of Babylonian wise men and shepherds in ancient Judaea, the unaided human eye was the only observational resource. Today binoculars give us more reach than was available to famous scholars up until the time of Galileo Galilei. Almost any commercially offered telescope sharpens our views to match those of Royal astronomers at yesteryear's Greenwich Observatory in England.

The most recent accumulator of starlight, the ordinary 35-millimeter camera, will record more than the naked human eye will ever be able to see. Together with a PROBLICOM discovery device (p. 110) or a STEBLICOM gadget (p. 112), which you can build yourself, photography can place even amateurs on the cutting edge of discovery.

Yet the most important resources at our disposal, the most acute sensors, are not the exotic highly vaunted CCDs (Charged Couple Devices) but rather our mind's eye coupled with our soul. In this area your own equipment may well outperform the world's leading observatories, can exceed in power the most sensitive man-made devices.

Intuitive feelings, serendipity and ordinary luck have all played important roles in the field of scientific discovery. But a camera in its case, just like the hook in a tacklebox, will never produce a 'catch'; only if it is properly 'baited' and 'trolled in the ocean' can it reward the patient and perservering 'fisherman' with a record prize.

The celestial firmament affords us the largest and deepest ocean. It presents us with the last and ultimate frontier. Like some vast universal liquid sandbox it is filled with countless sparkling jewels among which one and all can play. The admission is free.

Ben Mayer
Los Angeles
1984

PREFACE TO THE EVEN CHAPTERS

Astronomy, more than any other science, offers interested and reasonably intelligent people the opportunity to make truly meaningful and lasting contributions to the field. For example, did you know that:

The majority of novae – new stars – are discovered by amateurs, some of whom don't even own a telescope? Hundreds of non-professionals participate regularly in a worldwide program of star-brightness measurement? Most comets that reach naked-eye brightness are found by, and named after, amateur astronomers?

Professionals face the problem that in every square degree of the sky there exist dozens, if not hundreds or thousands, of stars, nebulae, galaxies and quasars, plus an occasional planet, satellite, comet or asteroid, many of which can be seen or photographed with inexpensive equipment. And there are 41 253 square degrees in the sky. There is no way that professional astronomers can keep track of them all. They need help.

This *Guide* is intended for lovers of astronomy – amateurs – who wish to do more than just look at the night sky and marvel at it all. If you own nothing more than a simple camera, or even if you possess a fabulously equipped observatory, this *Guide* will get you started. It explains in simple, non-mathematical terms how you can take positively stunning star photographs and then put them to use making valuable contributions to the science of astronomy.

For me, Ben's odd-numbered chapters provide a fascinating account, excitingly and humorously told, of how one very rank amateur got started and quickly progressed to become one of the world's best known, and in many ways most successful, amateur astronomer. In my even-numbered chapters you will find a more extensive commentary on much of what Ben writes, plus some additional material which gives, I suppose, the professional point of view.

We are grateful to the many people who gave us their assistance along the way. I especially wish to thank Harvard University and Gonzalo Alcaíno, Director of the Instituto Isaac Newton, who provided me with expert typists, Joan Verity and Ximena Hederra. We are also indebted to Leif Robinson and to the patient and talented staff of our publisher.

W. Liller
Viña del Mar, Chile
1984

Fig. 1.1. The camera was securely clamped to a harness which was attached to the chimney of the house. The harness was rocksteady. The chimney stood firm, a part of the structure of the building. The house itself was planted firmly on its foundations. There is just one small problem for astrophotographers: the whole Earth itself turns and moves. Ursa Major. 4-minute exposure. 50-millimeter lens, f/1.4. High-speed Ektachrome slide-film, ISO 400.

CHAPTER 1

Transcendental astronomy

Don't buy a telescope – yet.

The main reason is that you don't need one to start on your exciting journey to the stars. What you need you probably already own: a camera and a pair of binoculars. If you also have a telephoto lens for your camera you are fully equipped; otherwise, to complete your observatory, buy a good long lens and a 'C' clamp to attach your camera to any firm base. (Tripods come later, usually with the telescope.)

When I was still a youngster and plans for the Palomar Observatory had just been announced, I asked my Father how learned or famous an astronomer would have to be to get to look through the planned 200-inch telescope. I was disappointed when told that hardly anyone would actually be *looking* through this gigantic instrument because it was too precious to be used for anything except photography. At the time, I thought it was sad that all that 'power' would be wasted. 'Looking' seemed much more fun. What I failed to understand was that there would be no ordinary picture-taking; instead, Palomar would be collecting starlight and recording it on film for all time.

There is magic there. Just think of it for a moment: the human eye, regardless of the power of any telescope, can see a dim star's light only dimly, and a nebula only nebulously. That is why so many visual star observers are disappointed when they find a galaxy which looks so beautiful in a picture and all they can actually discern through a telescope is, at best, a blur.

However, through long exposures a camera can allow light to accumulate on film and produce the spectacular images with which we are all familiar. Only in this way do objects which the eye can just barely see become easily visible in pictures. Such records can themselves be collected, compared and studied at leisure.

The word 'collecting' cannot be overstressed. It dramatizes the gathering of faint light. In terms of the history of astronomy – from Egyptian temples aligned with rising stars to the fifteenth-century astronomer Copernicus, who proclaimed the Sun to be the center of our Solar System – collecting starlight was made possible only in the most recent past.

In the 1830s the genius of Louis Jacques Daguerre, a French painter and the developer of the art of photography, yielded the first long-exposure pictures which stored light to produce daguerrotypes and, in turn, the first records of celestial objects.

Measured by the age of our Earth, humankind has only existed on this planet for the briefest of time. Think of Cleopatra's 'needles', the two obelisks formerly in Alexandria, Egypt, one of which is now in London, the other in New York. Picture their 20-meter red feldspar heights as representing the age of our planet Earth. Suppose you were to place on top of one of these 3000-year-old monoliths a silver dollar coin to which a postage stamp had been glued. Before you would stand an imaginary time-line of our planet's history. To help grasp the relative timespans at one brief glance: if the 20-meter pillar of stone engraved with hieroglyphs corresponds to the age of this planet, then the relative thickness of the coin on top represents the time that life has existed on the Earth's surface. The thickness of the postage stamp constitutes the period during which the first vertebrates roamed the land. The printed layer of ink on the face of the stamp depicts, finally, the age of *Homo sapiens*.

Stone-Age man, through inborn fear of the unknown, could only view the stars with wonder and the Sun with awe; whereas modern man, through methodical observations, found reasons and solutions for what went on in the heavens. In retrospect it is interesting to observe how wrong some of these solutions were in terms of what we now know. Similarly, what we regard as the last and final truth today may well be found laughable tomorrow. The only thing that will have a lasting and unquestionable value will be the records which we keep. The Harvard University Collection of Celestial Photographs, the largest in the world, is preserved in a special building for astronomers to check back, make comparisons, and evaluate their own findings in the light of what has gone before.

Astrophotography does not need fancy equipment, nor does it require a thorough knowledge of astronomy to start. It calls for no more than a camera which can make time-exposures, a cable-release, fast black-and-white or color film, and a firm, steady base to which to attach the camera. A tripod is only one example of such a base. I started by clamping my 35-millimeter camera to a bolt on the stone chimney on my roof. The assembly seemed rocksteady but there was a problem – the stars seemed to move (Fig. 1.1).

It is easy today to say 'the stars seemed to move'. To early man the stars *did* move, and so did the planets, the Moon and the Sun. If modern man did not have the benefit of history's cumulative genius to draw upon, an individual without access to prior learning would instantly revert to the darkest beginnings of time when seeing was believing. Thus a solar eclipse would still be an inexplicable event which would put fear into our hearts.

One may wonder what Galileo could or would have done if he had had access to a roll of high-speed film and a camera to attach to his telescope, let alone a place where he could have had his film processed into color prints. But even without these, the legacy of the seventeenth-century Italian – his brilliant mind tormented by the Inquisition – will survive as long as humankind.

Imagine placing a large-aperture reflector telescope at the disposal of Isaac Newton, or showing him the revelation which modern spectroscopy has produced, all based on his discovery that the prism separates white light into a rainbow of colors.

Improvise in your mind a scene with Babylonian wise men in the year 2500 BC, or Mayan priests 3000 years later, having access to a modern digital clock or any other timekeeping device to aid them in their studies and observations.

Finally, imagine placing into the hands of today's scientists only one of the devices that will be available just 500 years hence.

Barring the thought of an atomic holocaust, it is fascinating to contemplate anything 500 years into the future. The Earth will almost certainly still be here. The chances of a cataclysmic collision with another celestial body is as remote as the chance of any six tennis balls, floating weightlessly in a hollow sphere the size of our Earth itself, coming into contact with any other.

Our life-giving 5-billion-year-old Sun (1 billion = 1000 million) is expected to have another 5 billion years of virtually unchanged existence before it will expand and raise the temperature on Earth past boiling point, eventually vaporizing our planet.

Thus, in the timetable of astronomy, the next 500 years are merely an insignificant, fleeting moment. When measured in human lifetimes or generations, however, five centuries represent the entire span of modern history, from the Dark Ages to men walking on the Moon.

In astronomy one becomes involved with endlessly long timespans and with even greater distances. Once you grasp the elegantly simple meaning of the term 'light year', the shackles of disproportionately small, earthly measurements will fall away.

In the same way that the sound of thunder, caused by the release of static electricity from cloud masses, takes time to reach our ears, so the seemingly instant flash of distant lightning, the visible manifestation of the electrical discharge itself, also takes time to reach our eyes. While sound travels about one-third of a kilometer (one-fifth of a mile) in a second, light covers a mind-boggling 300 000 kilometers (186 000 miles) in the same period of time.

One light year is quite simply the distance over which light travels in one year. In one minute, light travels 18 million kilometers. In one hour it journeys 60 times as far again. Multiply by 24 for one day and then by 365 for one year. That makes a *distance* of 9 460 800 000 000 kilometers per year. The measure is beyond comprehension. Like so many of the truths in astronomy, it 'passeth understanding'.

My Mother died a few months after I had embarked on my adventures with the stars. On the evening of her death, I walked out into the night and searched the November sky for the constellation of the Ram and within it the brightest star called 'Hamal'. I had found it in a list of bright stars which also told me that its designation was Alpha Arietis, that it had a magnitude of 2.2 and that it was 75 light years away from us. It was the distance which made me seek out this particular object in a constellation that I did not yet know. Only careful reference to a star chart by dimmed flashlight helped me pinpoint my star. When I finally saw it through binoculars, I was overcome by the awareness that the starlight I was seeing at that very moment had embarked on its journey to me at about the time my Mother was born and had been underway to my eye for the entire span of my Mother's life.

Hereafter, if someone on a planet near Hamal has a very powerful telescope trained on planet Earth, they might follow her life, from childhood, through adolescence to old age. All the time such heavenly observers would be viewing images from another time, pictures of the past, just as I was looking at 75-year-old radiance.

Observing stars and collecting their ancient light on film can become a passion. It leads one to look past our Solar System to the stars in our Galaxy and then out to where countless other island universes beckon. It is when we sense without properly seeing, when we feel without ever being able to reach, that astronomy rewards us with new awareness and fulfilment. It is then, in the face of the majesty and the overwhelming vastness of the Cosmos, that the beginning amateur and the learned astronomer become as one.

The comparatively greater knowledge of professional scientists still remains dwarfed by all they do not and will never know. The possibility even exists that beginners, due to their general ignorance of the limitations in the field, can make important contributions to science, even breakthroughs. All that is needed is devotion to the task at hand combined with a certain measure of discipline (see Chapter 7). The amateur need only bring with him the sense of adventure that goes with the exploration of uncharted paths, which are open to all from their rooftops or their backyards.

The word 'amateur' stems from the Latin word *amator* which means 'lover'. Perhaps we should be talking more about love and less about raw knowledge. In some instances, ignorance may even be bliss. Not knowing any boundaries may encourage us to embark on bold undertakings. Dogma will often stifle inventiveness.

The very fact that we are living on this Earth in this particular split-second of time – time without end – makes me regard the stars with profound feelings which border on what might be called a religious experience. It helps if one is aware of one's own transitory presence on this planet at this particular moment in the seemingly endless and immeasureable time called eternity.

The story is told of Heinrich Schliemann, the nineteenth-century German businessman turned archeologist who, on his own, undertook excavations to discover the ancient city of Troy. It is said that he was once asked by his son just how long eternity was. At the time, father and son were in Greece on a terrace overlooking Athens, with Piraeus, the harbor town, shrouded in mist some 3 kilometers away. 'If there were a wall of marble, ten yards high and ten yards wide, stretching from here all the way to Piraeus', the elder Schliemann motioned to the south, 'and a bird came along once every thousand years, drawing a silken scarf along the entire length of this gigantic stone, then, by the time the marble would have worn down to the thickness of a sliver, one second of eternity would have passed' (Fig. 1.2).

Fig. 1.2. 'If a bird came along once every thousand years, drawing a silken scarf along the entire length of this gigantic wall of stone, then by the time the marble would have worn down to the thickness of a sliver, one second of eternity would have passed.'

Fig. 2.1. A spiral galaxy, NGC 253, in Sculptor.

CHAPTER 2
First things first: fundamental thoughts and concepts

Was there a single most important discovery in the history of astronomy? Probably not, but most astronomers and historians of science rank highly the realization by Nicolai Copernicus that the Earth moved in an orbit about the Sun and was not located at the center of the Solar System. In fact many scholars mark the beginning of the Renaissance with Copernicus' great revelation. Coming nearly 100 years after the discovery of the New World, this great work was remarkable not entirely for its startling disclosures; after all, some of the pre-Christian Greek philosophers and scientists had discussed the same ideas centuries before. Moreover, in a certain very small sense, the Sun does revolve around the Earth. To be exact, the Earth and the Sun both revolve around their common center of gravity which happens to be inside the Sun and only about 450 kilometers this side of the Sun's center, a miniscule distance compared to the distance of the Sun from the Earth, 149 597 870 kilometers (about 93 million miles) (Fig. 2.2). The existence of all the other planets makes it virtually impossible to say exactly where the center of gravity of the Solar System is at any moment since each contributes its small gravitational pull on the Sun, as does every satellite, asteroid, comet and meteoroid -- and every inhabitant of the above.

Much of the greatness of Copernicus lay in his courage to suggest such an idea, or to propose any new theory at all. He lived, remember, during the Dark Ages and at the time of the Inquisition. Who was going to believe a little-known Polish astronomer who dared dispute the great Aristotle whose theories had dominated so much of science for more than 1800 years? But eventually it was realized that this relocation of the Sun and Earth removed many (though not all) of the cumbersome epicycles – those little circles on big circles – needed to explain why each of the planets moved in such strange ways relative to the background of more distant stars. Johannes Kepler added a much needed refinement 66 years later when he found that the orbits of the planets must be ever so slightly elliptical in shape, not circular, a fine subtlety which can best be appreciated if you compare an accurately drawn ellipse, having the same eccentricity as Mercury's orbit, with a perfect circle (Fig. 2.3). Of the major planets, only Pluto has a more eccentric orbit.

All of Kepler's work had appeared by 1618, only 9 years after Galileo Galilei first used the newly invented telescope to look critically at the stars, planets, Moon and Sun. The 'heretic' aspect of Galileo's many findings is well known. He too had courage, more than Copernicus since he chose to face the Inquisition and not to wait until he was on his death bed to publish his discoveries as did Copernicus.

Possibly Galileo's discoveries were the greatest. Or perhaps the credit should go to the developer of the telescope, whoever he was. Some believe it was an obscure Dutch optician by the name of Lippershey.

But the first person to provide explanations for many of the mysteries of planetary motion was Isaac Newton who, along the way, had to invent a new field of mathematics (calculus) for the purpose. Newton not only discovered gravity, which he did as a result of pondering Kepler's new laws of planetary motion, but he

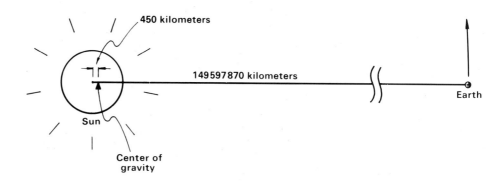

Fig. 2.2. If connected by a weightless rod, the Sun and the Earth would balance at the center of gravity located some 450 kilometers from the Sun's center. As shown here, the diameter of the Earth is about ten times larger than it should be relative to the Sun.

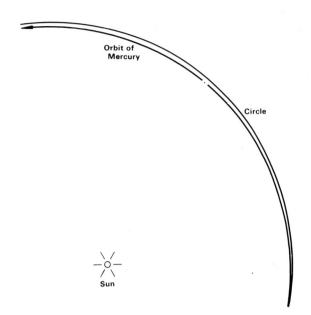

Fig. 2.3. The shape of the orbit of Mercury differs very little from a perfect circle, but the Sun is located far from the center. In this drawing the size of the Sun is correct in proportion to that of the orbit, but the dot representing Mercury is 40 times too large.

eclipses to notice the stars themselves. William Herschel, a German organist who fled to England to avoid military service, was one of the few who was more fascinated by the space beyond the planets than by celestial mathematics. (However, while making his careful stellar observations, he by chance discovered the planet Uranus, an event which brought him knighthood and a royal pension for life. Accidents will happen.)

Herschel built his own reflecting telescopes with sizes (mirror diameters) up to 18 inches, and used them to discover thousands of objects of interest that lay far beyond the Solar System – double star systems, star clusters, nebulae and galaxies. He was unable to tell what a gaseous nebula was made of or that a galaxy was an island universe of stars, but he did note their existence and, perhaps for the first time, pointed out the Milky Way as something significant in our part of the Universe.

Then came Albert Einstein. Many consider him the equal of Newton; both were deeply concerned about the fundamental properties of mass, motion, time and space. Einstein's theory of relativity added refinements to the theories of gravity and planetary motion and also, of course, showed that mass and

virtually invented physics since he had to work out the motion of bodies under various forces. Newton had other talents. For example, he wrote the first knowledgeable book on optics that explained why Galileo's telescope worked, how a glass prism could produce a whole spectrum of colors, and why a parabolic mirror could be used to make a telescope, a better one, in fact, than one made wholly of lenses.

Newton lived to a ripe old age, but it is interesting to note that after he reached 30, his scientific output went essentially to zero. For a while he lectured at Cambridge University, but then he went to work as Her Majesty's Keeper of the Royal Mint. While one might consider this quite a come-down, it no doubt paid well. This gave him time to dabble in his later passion, theology.

The seventeenth century, during which all the immense contributions of Galileo, Kepler and Newton emerged, was unquestionably a fantastic era for astronomy. What about the eighteenth, nineteenth and twentieth centuries? With few exceptions, until the beginning of this century, astronomers were so overwhelmed by Newton's work that little else was added. Everyone seemed too busy calculating orbits and predicting

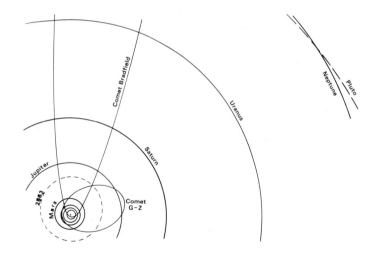

Fig. 2.4. Mercury, Venus and the Earth most enjoy the heat of the Sun; their orbits are not labelled. Comet G–Z (for Giacobini–Zinner) has a period of 6.5 years and is typical of many short-period comets. Comet Bradfield was discovered in December 1980 and reached magnitude 2.5 shortly afterwards. It returns to our neighborhood once every 29 000 years. Note that from 1980 until 1999 Pluto will be closer than Neptune to the Sun. All objects shown here, except Comet Bradfield, revolve around the Sun counterclockwise as seen from the north. As for Asteroid No. 2863, read Chapter 8.

energy were one and the same thing, only in different forms. We now know how to calculate the quantity of energy (E) possessed by a piece of mass (m); they are related simply by the square of the speed of light (c): $E = mc^2$.

Fig. 2.5. *The 200-inch (5-meter) telescope on Mount Palomar. This 50-tonne monster is so delicately balanced that a one-twelfth horsepower motor is sufficient to drive it westward as it tracks the motions of the stars, planets and galaxies.*

Fig. 2.6. The Multiple Mirror Telescope of the University of Arizona and the Smithsonian Institution on the summit of Mt Hopkins, Arizona. The six 1.83-meter mirrors have a collecting area equivalent to a single mirror with a diameter of 4.5 meters, making it effectively one of the largest telescopes in the world.

While Einstein was a thinker, a theoretician, many of the astronomical discoveries of recent years have been in the area of observational science. Four major developments have played an important role in the development of modern astronomy and the closely related field of astrophysics:

1. *Photography*. John Herschel, the son of the discoverer of Uranus, was one of the first to experiment with the permanent recording of light, but it took many decades for the technique to develop from an amusing curiosity in the early nineteenth century to a straightforward and important method of making images. Moreover, photography was a way of recording stars and nebulae much fainter than the eye could see (because long time-exposures were possible). Wavelengths that the eye could not see (the ultraviolet and the infrared) became visible and records were permanent and absolutely non-subjective. In recent years television systems and microchip technology, aided enormously by computers, have grown tremendously in importance. Still, photography continues to be the 'bread-and -butter' detector of astronomers.

2. *Giant telescopes*. The construction of the 100-inch reflector at Mount Wilson in 1917 inaugurated a new era of astronomy. Even so there were few additions to the inventory of large telescopes until the 200-inch (Fig. 2.5) was finished in 1947. More recently, twin 4-meter telescopes have been put into operation in Arizona and in Chile, and telescopes nearly as large scrutinize celestial bodies from mountaintops in Hawaii and the Canary Islands. As of now the world's largest telescope, a 6-meter reflector, scans the skies in the Soviet Union. Besides its size, this monster is unusual in that it uses an alt-azimuth mounting, unlike the tipped-over type of mounting used by nearly all other large telescopes. The multi-mirror telescope in Arizona (Fig. 2.6), composed of a cluster of six 72-inch mirrors all focused together to give the effect of one 176-inch reflector, also uses an alt-azimuth mounting. Perhaps all the large telescopes of the future will be made in this way.

3. *Radioastronomy*. Karl Jansky, an engineer working for the Bell Laboratories, was the first to notice that weak radio signals came from the sky, mostly from the direction of the Milky Way. That was in 1932. Following World War II radioastronomy blossomed quickly into a major science. It developed first and most rapidly in Great Britain, the Netherlands and Australia. Now most of the industrialized nations of the world have major radioastronomy facilities. More and more often, those huge, familiar-looking parabolic radio receivers, which we simply call radiotelescopes (Fig. 2.8), are connected together electronically to form a worldwide network yielding many of the advantages of a single 12 700-kilometer-diameter telescope. With them we are learning fascinating details about quasars, pulsars and other strange objects discovered first by radioastronomers.

4. *Space astronomy*. At the Earth's surface we receive a distorted and incomplete view of the Universe. Many kinds of radiation – gamma-rays, X-rays, most of the ultraviolet and infrared, and some radio waves – are unable to penetrate the atmosphere, and much of what does get through is so distorted and weakened that ground-based telescopes never reach their full potential. Spacecraft-borne instruments have only recently begun to make the thousands of new discoveries that are ushering in the next great era of astronomy. At the time Sputnik went up in 1957, no one had any idea how the surfaces of the planets and satellites would look when viewed from close-up, or what the sky would reveal when viewed using X-rays and gamma-rays. In fact, the first star seen with an X-ray detector was found accidentally by a rocket experiment designed to measure the Moon's brightness in X-rays. (The Moon, it turned out, was too weak to be detected.) New and bigger space telescopes are being readied for orbit and we can only guess at what will be found in the next few years, both in our Solar System and beyond it.

Obviously, there have been many other people and many other developments that have brought us to where we are today in the field of astronomy. Harlow Shapley re-played the role of Copernicus in 1918 when he showed that the Solar System was not at the center of our spiral Milky Way Galaxy, but near one edge. Edwin Hubble used the 100-inch to demonstrate that the Universe is expanding. Ever since, many have pondered whether it will expand forever or will turn around and start contracting. And then what? What was there before expansion started in the first place? Are there other universes? We will go into more detail on these and related questions in the final chapter.

Astronomical distances

In astronomy we are continually dealing with immense distances, and in Chapter 1 the *light year* was introduced as evaluated. While this unit of *distance* (not time – a point often lost on the uninitiated) is frequently employed by professional astronomers, especially when discussing distances to the outer parts of the Universe, the more common unit used by professionals is the *parsec* the value of which is 3.26 light years. The definition of a parsec is simply stated: a person one parsec away from us would see the angle between the Earth and the Sun when the two are farthest apart as exactly one second of arc (a sixtieth of a sixtieth of a degree). The distance to Proxima Centauri, the star nearest to Earth, is 4.2 light

Fig. 2.7. The Hubble Space Telescope will have a primary mirror 2.4 meters in diameter and an impressive battery of instruments automatically available to the observer on Earth. Plans are to put HST in an 800-kilometer orbit within the next few years. This photograph of a model was supplied by the Space Telescope Science Institute, operated by the Association of Universities for Research in Astronomy, Inc., for the National Aeronautics and Space Administration.

years, or 1.30 parsecs. The distance to the nearest spiral galaxy, the Andromeda 'Nebula', is about 730 000 parsecs or nearly 2.5 million light years. (This spiral system is the most distant object that can be seen with the naked eye.)

For measuring distances within our Solar System, miniscule compared to a light year or parsec, the standard unit of length is the average distance of the Earth from the Sun, 149 597 870 kilometers (92 956 000 miles), called the *astronomical unit*. The Earth–Sun distance varies, due to the ellipticity of the Earth's orbit, by a little under 5 million kilometers, which puts us 1.017 astronomical units away from the Sun at aphelion, our most distant point, and 0.983 astronomical units away when we are at perihelion, our nearest position. While it does not feel like it in the Northern Hemisphere, we are closest to the Sun in early January.

In astronomy one never realizes fully how incomprehensibly large this Universe of ours happens to be. Even the size of our own little Solar System is fantastic, and oft-quoted examples – such as it would take a high-powered rifle bullet, travelling at 16 000 kilometers per hour (Mach 13) nearly 5 years to reach Jupiter – leave minds benumbed, even those of professional astronomers.

The magnitude system

No one knows who was the first person interested enough in the stars to record their brightnesses. Beginning with vague statements like 'a star of the first magnitude' which literally meant a star whose brightness category was No. 1, there evolved a precise system of definition which carried over this ancient terminology. Today, astronomers are able to measure brightness with an accuracy of a small fraction of a magnitude, but they still use the old words. In Chapter 20 we will describe how one can measure brightnesses with astonishing precision by using a simple photo-electric photometer.

Any good star catalog or atlas provides magnitudes of stars, planets and other objects of interest. Practise estimating magnitudes so that you can quickly judge brightnesses of astronomical objects seen with your own eyes, your camera, and eventually your telescope. In Chapter 8, we will talk about stars of variable brightness, and you can soon be measuring their magnitudes as a part of an international program.

We should make it clear, here and now, that a *difference* of one magnitude corresponds to a *ratio* of brightness of just over 2.5 times. A two-magnitude difference tells us that one object is 6.3 times brighter than another ($2.5 \times 2.5 = 6.3$). A difference of three magnitudes defines a brightness ratio of 16:1. Four magnitudes corresponds to 40 times; and five magnitudes, exactly (by definition) 100 times. Always remember that differences in the magnitude system correspond to ratios on a brightness scale. The reason why astronomers use what seems like a weird system of measuring star brightnesses is simple: the eye–brain combination works that way. Brightness ratios are perceived as differences; the difference in brightness between stars of magnitude 2 and 4 *seems* to be twice the difference in brightness between stars of magnitude 2 and 3. Notice also that the *larger* the magnitude, the *fainter* the star. Like a golf game: the lower the score, the better the game.

Highly precise magnitude standards have been established all over the sky by modern-day photometers attached to large telescopes. Vega, the bright star nearly overhead in the early evening as seen from temperate northern latitudes in mid-summer, is one of the primary standards and shines brighter than first magnitude. To be more exact, its visual magnitude is $+0.04$. The brightest star in the Southern Cross has a magnitude of $+0.85$, while the stars in Ursa Major (the Great Bear, Big Dipper, or Plough) average about 2.0. To relate these quantities to earthbound values, astronomers have determined that a candle flame has a magnitude of exactly 0.0 at 2058 feet (0.63 kilometers). The same candle seen from 6.3 kilometers will appear to have a magnitude of $+5$. At 13.1 kilometers it will be at the limit of the naked eye visibility at magnitude $+6.6$.

Objects much brighter than Vega will have negative magnitudes, a detail that the ancients ignored by calling everything that was bright 'first magnitude'. The brightest star in the sky, Sirius, known to the ancient Egyptians as the Dog-star, has a visual magnitude of -1.47; the planets Mars and Jupiter, when at their brightest, are magnitude -2.5; and Venus, the brightest planet of all, reaches -4.6. Continuing onwards, astronomers have found the magnitude of the Full Moon to be -12.7 and that of the Sun to be -26.8. At the other end of the scale, with an average pair of binoculars you can usually reach to ninth or tenth magnitude. With an 8-inch telescope your visual magnitude limit becomes about 13, although a long-exposure photograph with the same telescope can show images to eighteenth or nineteenth

Fig. 2.8. (overleaf, pp. 18–19) Only a few of the 27 parabolic antennae making up the Very Large Array of radiotelescopes in New Mexico are visible in this photograph taken with a double rainbow in the background. Each dish is pointable and can accurately track a radio source as it moves with the rest of the stars. Each of the 25-meter-diameter telescopes can be moved along precision railroad tracks measuring 21 kilometers in length, thereby making it possible to 'see' at radio frequencies much more clearly than with a single antenna. This fantastic instrument is a part of the U.S. National Radio Astronomy Observatory, operated by Associated Universities, Inc. under contract with the National Science Foundation.

magnitude. The faintest star that has so far been recorded has a magnitude of about +25, over 10 billion times as faint as the star Vega. Halley's Comet, when first detected in 1982 on its voyage in towards the Sun, was estimated to be magnitude 24.2

The full brightness range – the 52 magnitudes from Sun to faintest star – corresponds to a brightness ration of 630 000 000 000 000 000 000 times, or in the short-hand notation of the scientist, 63×10^{19} times.

Magnitudes can also be used to measure colors. The eye is most sensitive to yellow and green light while the best photographic materials in use at the turn of the century were mainly blue- and violet-sensitive. On early photographs blue stars were brighter than they appeared to the eye, and the terms *photographic magnitude* and *visual magnitude* were employed to indicate the different scales. We call these two kinds of magnitudes B and V (for 'blue' and 'visual'), and we define the *color index* of a star as simply 'B minus V'. Vega has a $B-V$ color index of zero, by definition. A bluer star will have a negative $B-V$, meaning its blue magnitude is less (brighter) than its visual magnitude. Yellow stars, like Arcturus and Pollux, have color indices which are slightly greater than +1. Very red stars, like Antares and Betelgeuse, have a $B-V$ near +2; the $B-V$ of the Sun is +0.63.

Finally, let us introduce one more very important concept regarding magnitude, and that is *absolute magnitude*. When one compares Vega at an apparent visual magnitude $V = +0.04$ and Polaris at $V = +2.1$, what is meant of course is that Vega *appears* to be 6.7 times brighter than Polaris. However, it so happens that Vega is much closer to Earth (26 light years away) than is Polaris (650 light years away), and Polaris is, in a real sense, the more luminous object. The absolute magnitude of a star defines how bright the star would appear if it could be moved to a standard distance of 10 parsecs, or 32.6 light years. The absolute visual magnitudes of Vega and Polaris are +0.5 and −4.4, respectively. If Polaris were 10 parsecs away, it would outshine all other stars in the sky and all but one of the planets. The Sun, whose absolute visual magnitude is +4.8, would be an inconspicuous star in the sky if it were 10 parsecs away. The Earth would be an insignificant twenty-eighth-magnitude object, undetectable in any instrument yet devised by man, except for Space Telescope.

Fig. 2.9. The Large Magellanic Cloud was once probably a spiral galaxy but because of its proximity to our own very much larger Galaxy, it has been badly warped out of shape due to our gravity. The LMC, together with its smaller companion the Small Magellanic Cloud, is the closest galaxy to us earthlings. Photo taken by the author from his backyard with a 135-millimeter lens and hypersensitized 2415 film. (2-minute exposure.)

Fig. 2.10. The gossamer giant radiotelescope antenna at Goldstone in the Californian desert.

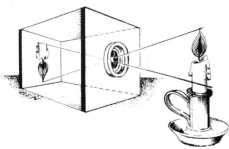

Fig. 3.1. Simple cameras are identical to the early camera obscura *which, literally translated from the Latin, means 'obscure chamber' or 'darkened box'.*

Fig. 3.2. (above) Constellation Taurus, 15-second exposure. 50-millimeter lens, f/1.4. High-speed Ektachrome slide-film, ISO 400.

CHAPTER 3

The patient eye — the camera

I have to make a confession: I *did* buy a telescope right at the start. The purpose for my purchase was not celestial or astronomically scientific at all. The heavenly bodies which I had planned to watch were of the bikini-clad, Californian variety. The view from my house on the hill includes a dozen swimming pools and affords a multitude of spectacular vistas.

An old saying tells that the difference between men and boys is the cost of their toys. Thus justified, I invested rather heavily in a beautiful reflector, the kind of telescope where, in place of a lens system, a large curved mirror collects the light, then reflects it to be focused for the eye or camera to see.

The cool professional feel of the precision-made instrument, the smooth action as I swung it about in its ball-bearing base, the phenomenal and enticing mirror images visible through the squat barrel were all sources of aesthetic delight to me. Looking at terrestrial objects through the telescope in the heat of a summer's day provided me with my first lesson concerning the importance of good viewing conditions: the air currents caused by the hot air rising from the landscape made everything seem to swim in clearly visible turbulence. People and objects perhaps only a mile away seemed to turn to jello when viewed at high magnification. Only lowest magnification permitted a steady view but, at the same time, presented a much larger field image and much less apparent detail. Basic applied optics! Higher magnification is possible only when atmospheric conditions permit. This applies to night viewing also.

Obviously there seemed to be a trade-off between the detail I would be able to resolve and the size of the object itself, between how much I would be able to see and how well one would be able to see it. Indeed, the laws of optics dictate that a viewer must forego magnification or 'power' for resolution and clarity.

With the approach of fall, much of the novelty of starlet-gazing had gone with summer. By October I had turned my telescope skyward and started on my astronomical journeys. That is how I began my romance with the stars. Almost from the beginning, much of my joy came from astrophotography. I soon found that an old 35-millimeter camera and its telephoto lens could be mounted 'piggyback' on the telescope with a simple clamp which manufacturers provide for just this purpose. This offers the chance to photograph objects easily, once they have been pinpointed with the telescope and viewed. The telescope support will serve as a tripod with all systems pointed in the same direction.

An important observation should be made at this point: the simpler the camera, the easier the photography where the night sky is concerned. Standard 35-millimeter cameras are usually equipped with 40-, 50- or 55-millimeter lenses. These produce pictures which approximately record the 'area' or field of view as seen by the naked eye. Such lenses have a variety of light-gathering capabilities, commonly referred to as 'speed'. Although they are all designed primarily for standard terrestrial photography, most are also ideally suited for night-time celestial picture-taking. It matters little what kind of camera one owns, for we shall quickly find out how much better suited is any camera for seeing what is in the night sky, than is the unaided human eye.

Even when the eye is properly adapted to the dark — you have to wait in the darkness for a while before you can properly see the stars — we can only detect starlight, but never collect it. The film in a camera, on the other hand, patiently and slowly gathers even the faintest light and stores it in the sensitized emulsion, so that we can observe it at our leisure after development: 30 seconds' worth of light, or even 10 minutes of exposure.

Collecting starlight or 'photons' is a little like putting a dish out in the open when there is only the merest drizzle of rain. It may take a little while, but after some time has elapsed the dish will be partly full of water, eventually full to the brim.

Let us briefly examine the simplest camera for a quick overview. In essence the principles governing the 'pinhole in a shoebox' camera will always apply. The basic configuration will show light entering through a small hole into a darkened box. The pinhole aperture takes the place of a lens, admitting the light rays and projecting them upside-down on the far inside of the shoebox. This surface is called the 'focal plane' which may sound impressive. It merely defines the area on which an image is focused (see Fig. 3.1).

One can observe the upside-down image if the box is large enough to get inside. If frosted glass is placed in the focal plane it is possible to observe it from outside the box. Such simple cameras are identical to the early *camera obscura* which, literally translated from the Latin, means 'obscure chamber' or 'darkened box'.

Much progress has been made in camera design and efficiency since the old *camera obscura* first bewitched audiences in darkened rooms at country fairs or turn-of-the-century seaside resorts. Dramatic changes have been introduced since photographers with cumbersome concertina-shaped boxes peered out from under black cloths and asked our great-grandparents to 'Watch the birdie!'. Continuous improvements were made in the entire photographic process and especially in the sensitivity of the emulsions used to record light. Where early so-called daguerrotypes would take minutes to expose, often yielding blurred pictures which in time would yellow and fade, today we can arrest the motion of a bullet in flight – in color, if we choose.

'Faster' high-speed films with remarkable light sensitivity resulted in ever-decreasing exposure times needed to record photographic images which, for us, can include the most distant stars and other faint celestial bodies. Improvements in the quality of lenses also benefited astrophotography, resulting in 'faster' optical systems.

Automobile speed is expressed in miles or kilometers traveled per hour. The relative speed of a camera lens is expressed in terms of the amount of light passing through in a certain period of time. A 'fast' lens is one which allows great amounts of light to pass in minimum time. A 'slow' lens may permit an equal amount of light to flow through, but it will need much more time for the light to transit.

Greatest optical speed is highly desirable simply because many photographic subjects move, or seem to move. I attached my camera–telescope combination to the chimney of my house so I could photograph from the flat roof. It was a rocksteady arrangement. The chimney did not move, being part of the house. The house stood firmly rooted to its foundations. But the Earth on which the house stood did move, which caused some early problems (see Fig. 1.1). It caused the stars to seem to be moving. For this reason it became imperative that I should take my star photos in the shortest possible time (see Fig. 3.2).

How fast is fast? Let's just say that the pinhole camera of yesteryear was agonizingly slow, even though the light only had to pass through air, and not through the glass of a lens. The bigger the opening which admits the light and the better the glass through which it travels (lens thickness and quality), the less time is needed to produce a picture. Speed in lenses, which results from a favorable combination of large aperture and good glass or crystal, is expressed as 'f' which can be said to stand for 'fastness'. The smaller the 'f' number (stop number) marked on the lens rim, the faster the speed. A lens with a low 'f' number such as f/1.4 is 'faster' than a lens with a rating of f/2.8. It is, in fact, four times as fast. It takes the same amount of light four times as long to pass through an f/2.8 lens as it does to move through an f/1.4 lens.

One can compare light traversing a lens to water flowing through a pipe. Just as the diameter of the pipe determines how much water will be able to stream through during a given period of time, the aperture of the lens controls how much light will pass and how quickly. If there are obstructions within even a large pipe, the amount of water flowing will be considerably slowed, just as poor glass in inefficient lenses will slow the transmission of light. Remember:

BIG PIPE = LOTS OF LIGHT PASSING THROUGH

LITTLE PIPE = LESS LIGHT.

It is precisely this fact which determines that a large-diameter telescope or lens can easily allow us to see (and to magnify) or photograph, what is simply impossible for a small-aperture system to detect.

Since the terms 'power' and 'magnification' both refer to the scale or ratio by which an image is increased in size through an optical process we can explode once and for all the advertising myth of 'telescope power'. This term has often been used by unscrupulous makers and sellers of telescopes to promote third-rate instruments which lack both large diameters and high speed. It should be remembered that it is not possible to magnify (by any power) any image which has not first been clearly resolved by an optical device of sufficient diameter (aperture).

Lens aperture, therefore, is really the most important single factor in the selection not only of telescopes but also of cameras for astrophotography. When you observe the human eye, you will note that nature follows this rule and that the dark-adapted eye, which can see best under night conditions, is a 'large-aperture eye' where the pupil is wide open or 'dilated'. It may take our eyes some 10 minutes to dilate to where we get the best use for star observation.

Another name for aperture is 'light-gathering power'. A reflector telescope with a diameter of 5 inches can easily have 200 times the light-gathering ability of the human eye, while a 14-inch instrument will exceed the eye by a factor of almost 2000. To see a planet as well as a telescope-aided viewer does in one second, the unaided eye would have to do the impossible: absorb 1000 seconds' worth of light in one second.

The advantage of increasing aperture size, to allow more light to enter an instrument, is the reason for the construction of ever larger telescopes such as the Palomar 200-inch reflector or the larger multi-mirror telescopes already in use or in the planning stages. The placing of a large telescope in space, above the obscuring dust and moisture surrounding our planet, allows for the most efficient observing.

Little wonder that the major observatories of the world pride themselves – never on any magnifying powers – merely on their diameters. The 200-inch Palomar telescope used to be the biggest eye on the sky in the Northern Hemisphere, but it has now been overtaken by a Soviet-built 236-inch giant in Zelenchukskaya in the Caucasus Mountains. It was constructed at an altitude of about 2000 meters. In this observatory a mirror weighing 42 tonnes and 6 meters in diameter assists scientists to penetrate ever deeper into distant space. There are also multiple-mirror telescopes which are larger still.

To return to our simple cameras, let us establish at the outset that while the diameter of a Pentax, a Canon, Nikon or Miranda lens cannot begin to approach that of any telescope, the lenses which are available here are of high quality usually with very high speeds, and are ideally suited for star photography. Low f values are common such as f/1.4, f/1.8 or f/2.

Unfortunately the standard lenses have relatively short 'focal lengths'. The magnifying power of a lens and the size of the image recorded relate directly to the 'focal length' of the lens. This length could be compared to the distance between the pinhole on the one side of the shoebox and the surface (focal plane) onto which the image is projected (the focal length thus equals the length of the shoebox). Focal length in lenses determines basic magnifying power which is why so-called 'powerful' telescopes or binoculars are always physically longer than their regular counterparts. It is easy to remember that a 500-millimeter lens will bring subjects much closer than a 135-millimeter lens.

The lens which is attached to the television camera that focuses on the intense expression of the quarterback in mid-field and transmits it to your home to fill every inch of the screen with helmeted determination, looks like a stove pipe and may be one meter or more in length. By comparison, the wide-angle shot from the overhead dirigible which displays the entire stadium together with the parking lot, is taken with a lens that is no more than a few centimeters long.

One simplifying aspect of celestial photography with basic cameras is that no focusing is required when working with lenses. The infinity setting, ∞, which is often marked on lenses in red, is the only setting to use. Stars, comets, meteors are all infinitely far away and even the Moon can be photographed at this setting.

The next most important questions concern the length of exposure. Clearly there is no way to compare night photography to daylight snapshooting and built-in exposure meters will be of no use whatsoever. There are two reasons for this: (1) it is too dark to take readings or even to see the meter needle if it should move; (2) in most cameras the 'B' setting used for extended time-exposures deactivates the photoelectric measuring device in the camera, so that even in daylight no readings could be taken in the 'bulb' setting.

'B' does actually stand for 'bulb'. This term dates back to the days when photographers pressed a rubber bulb in their hands, manually holding open the shutter on lenses which were activated by air pressure. The lens stayed open so long as the ball was being pressed. The modern cable-release has taken the place of the rubber bulb and hose, allowing the opening and closing of a shutter without transmitting vibrations from the hand of the photographer to the camera.

Unless the subject is the Full Moon, or one shoots the brighter planets, no night exposure will ever be shorter than one-quarter of a second in length. For this reason, all the costly timing features which are found on cameras today become unnecessary for our purposes. The one-thousandth of a second exposure capability of a costly camera is just as useless as the synchronized flash terminal, the built-in light meter or the hot-shoe. If you buy a camera for astrophotography, the simplest one will do just fine. You can even buy one in a pawnshop. Put your money into the best lens you can afford, the fastest you can get (low f number): a standard 50-millimeter; a medium-long one (135 millimeters) is a good second lens. Lens speed is the most important thing for which to look. Fast longer lenses cost a great deal of money and once they get past 135 millimeters, high speeds are hard to find. A 135-millimeter telelens with f/2.5 can cost twice as much as a seemingly identical f/4.

The speedier f/2.8 lens is almost twice as fast as its f/4 equivalent. Thus, an exposure which takes 10 seconds to make with the slower lens, needs only 5 seconds with the f/2.8. 'Big deal', you may say, but once you try for longer exposures the difference between a 5-minute shot and one taking 10 minutes will become critical. Don't forget, the Earth keeps moving.

Take my word for it, speed is the name of the game, worth every extra penny.

Fig. 4.1. This globular cluster, called Omega Centauri, contains nearly a million stars all believed to be 15 billion to 20 billion years old. It is one of 131 known to belong to our Galaxy, but there are probably 50 or so others hidden behind the giant dust clouds in interstellar space.

CHAPTER 4

More about cameras — telescopes too

Stripped to their essential optical components, cameras, binoculars and refracting telescopes need only two parts to operate properly: one lens to focus the light, and a second lens to magnify the focused light. (In a camera, the second lens will be the projector lens or magnifier used to examine the developed film.) While there now exist many lens designs, often immensely complicated, many extremely simple systems can be used to make remarkably effective astronomical instruments. Moreover, there is a better way to focus light than with a lens, namely with a mirror. To start with let us examine the basics of cameras and refracting telescopes; later we will discuss reflectors.

Refracting systems

Cameras, of course, never come equipped with an eyepiece, and a second lens is not absolutely essential to examine the camera's finished product, typically a black-and-white negative or a 35-millimeter color slide, but astronomical subjects usually require it. Of course that second lens could be the one in your slide projector and often is. A truly amazing amount of information can be stored on a single piece of film and if it is not examined carefully, a lot of information will go unretrieved. Astronomically, that could mean a missed comet, an undiscovered nova, or an unappreciated wisp of nebulosity at the edge of the Trifid Nebula.

With a refracting telescope or camera, the main lens, or *objective*, forms an image in mid-air at the end of the tube or inside the camera case. A camera captures the image on film; with a telescope, the image can be examined with an eyepiece (which is really nothing more than a high-quality magnifying glass) or recorded on film, thus turning the telescope into a special – and very powerful – kind of camera. At the focus of a really large telescope like the 200-inch Palomar, the image of a rich, compact globular star cluster seems to hang suspended in mid-air. One feels as if the several hundred thousand stars, all glittering brilliantly in a single plane, could be scooped up and taken home for later study. In a very real sense they can be, because their images do, in fact, lie in a single plane. If you were to put a piece of film, or better yet a photographic glass plate, precisely in that plane and expose it for 20 or 30 minutes you would have a spectacular photograph. You can gain a little appreciation of how this would appear by looking closely at Fig. 4.1, which is a photograph, taken by Ben Mayer, of Omega Centauri, one of the biggest and brightest globular clusters in the sky.

Except to make sure that the telescope is in focus and pointed at the right object, or to keep the telescope persistently tracking the slowly moving stars precisely, the professional astrophotographer rarely looks through a telescope; it is used exclusively as a camera. Conversely, if you put a small magnifier at the focus of your camera, you can make it into a telescope. Try it! If you happen to have a piece of ground glass in the house (a bit of waxed paper will also work, or you can make ground glass simply by rubbing a piece of window glass with a bit of fine-grade sandpaper), you can verify where the focal plane is by opening the camera back and placing the ground glass where the film goes. If you already own a telescope, you can do the same thing, only now you will notice that the focal plane is forward of the eyepiece position.

Binoculars were invented simply because it is difficult and uncomfortable for most people to keep one eye shut for long periods of time. Here we have two identical telescopes – spy glasses – rigidly held together and accurately aligned so that they will aim at the same subject together. There is usually a pair of multifaceted prisms inside, whose purpose is twofold: (1) to 'fold up' the light paths and make the entire package more compact; and (2) to bring the light which has been focused separately by each of the objective lenses to positions spaced apart by a distance equal to that between the eyes. Since eye separations vary, the binoculars have to hinge in a way to permit the eyepiece separation to be varied. Moreover, your two eyes may very possibly require different focus settings, and one of the eyepieces is usually independently focusable. Because of all the complications brought about by having these two identical minitelescopes operating precisely together, binoculars are (1) relatively expensive, and (2) easily knocked out of adjustment if dropped. It is both cheaper and simpler to buy a *monocular* and an eye-patch – or to train the muscles of one eye to keep it shut without discomfort.

If you do own binoculars or plan to buy a pair, you probably have noticed that they are rated by numbers like 6 × 30, or 8 × 40, or 12 × 50. The first number is the magnifying power; the second is the diameter of each objective lens in millimeters. If you use binoculars to look at objects in the daytime, the second number is unimportant. For night-time use, it is tremendously important, more so than the magnifying power. As Chapter 3 emphasized, the larger the lens diameter, the more light will be collected, and the fainter the objects you will be able to see, whether they be stars or ships in the night.

How do lenses focus light? First of all, instead of visualizing light as made up of waves, think of it as consisting of streams of super-small particles or packets of energy. The 'dual nature of light' is a concept that has been used by scientists for years. The name given to these light particles is *photons*. Consider the photons leaving a distant star; they go whizzing off in all directions traveling at the speed of light in a vacuum, 300 000 kilometers per second (186 000 miles per second). We are interested only in those few photons which are headed in one very specific direction, namely towards that speck in the sky, as seen from the star, which is outlined by the objective lens of our telescope or camera. These photons will be traveling in almost exactly parallel paths when they reach us, many light years away.

When these or any photons enter a piece of glass they immediately slow down. The speed of light is less in glass or in water than it is in air or in the near-perfect vacuum of outer space.

In general, the denser the refracting medium, the slower the light travels. In ordinary glass, the velocity of light is about a third less than it is in a vacuum.

What happens when photons enter a flat piece of glass that is tilted to the direction of their travel? Consider three photons that years ago left a distant star at the same time, traveling side by side (see Fig. 4.2). When the first photon enters the glass, it is immediately slowed down; a fraction of a microsecond later, the middle photon suffers the same fate; and then the last photon. The whole line of photons, once inside the glass, has swung around and altered its direction like musicians in a marching band going around a corner. The steeper the angle of the glass, the greater the change, and the more the light is *refracted*. If the glass is a window pane with both sides flat and parallel, just the opposite effect occurs when the light leaves. The end result is that while the light has been displaced, there has been no overall change in direction, which is why windows do not appreciably distort the transmitted image, even if looked through at a fairly steep angle.

If the flat piece of glass is now replaced by a lens-shaped piece (Fig. 4.3), the effect is different. Each photon is refracted by an amount depending on its distance from the center of the lens.

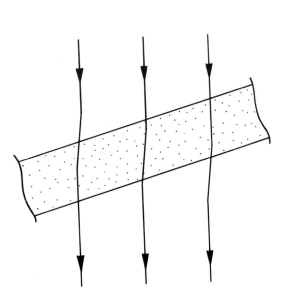

Fig 4.2. Light passing through a piece of plate glass may be slightly diverted but there is no distortion if the surfaces are perfectly flat. Glassed-in observatories do exist and make it possible to observe in full comfort.

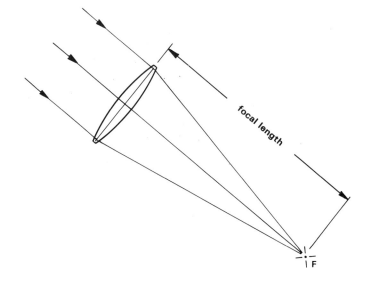

Fig. 4.3. Parallel rays of light such as those coming from a distant star are focused to a single point by a simple lens – or very nearly so. This lens has a focal ratio (focal length divided by diameter) of just over 2.4, or as they say in the camera ads, f/2.4.

The central photon entering head-on goes straight through, while the edge photons are diverted most, and all the light *focuses* finally at a single point which is the *image* of the star from which the photons originally came. A piece of film placed at the focus would record the once parallel-moving parade of photons as a single, sharp point.

Remember that central light rays go straight through a lens. (If they strike the surface at an angle, they will be slightly displaced laterally, but will not change direction.) With this information you can understand the workings of a camera (or a telescope without the eyepiece). You can now understand better why the scene appeared upside-down when you looked at the focal plane of your camera (or telescope).

Unfortunately, a single simple lens does not function perfectly. The biggest problem is color. Each hue focuses at a different distance from the lens. Violet light, having the shortest wavelength of visible light, is refracted most. It has therefore the shortest focal length, and comes to a focus nearest the objective lens. Blue is next in order; then green, yellow and orange; the red image lies farthest from the lens. Each color of the rainbow has a different focus point. However, by adding one additional lens we can greatly reduce these problems. We then have what is called an *achromatic lens*.

In Fig. 4.4 we show a two-element objective, which solves the color problem adequately for many purposes. A simple achromatic objective uses two types of glass, one having a strong coloring effect for a given focal length compared to the other.

Therefore, the focusing errors introduced by the main double convex lens are largely eliminated by the accompanying plano-concave lens. In actual practice, the two-element lens still does not totally cancel out the aberrations of the first lens, but with proper design the blue and red images can be brought together at the same focal point with the light of all the other colors coming to a focus nearby.

If one wants light of several wavelengths to be correctly focused, more components must be included. A good focus over a wide field requires still more lenses, sometimes curved into bizarre shapes and made of exotic materials. The expense increases rapidly with lens size and desired degree of correction.

Look at Fig. 4.3 again. It is important to understand what is meant by *focal scale*. Assume that a comet's tail is as long as Orion is tall, about 20 degrees in the sky. If the length of the tail image on the focal plane is 10 millimeters, then the focal scale is 20 degrees per 10 millimeters, or 2 degrees per millimeter. You can also calculate the focal scale in degrees per millimeter of your own camera quite simply: divide its focal length (in millimeters) into 57.3. Many camera lenses have focal lengths of around 50 millimeters; the focal scale will be just a little over one degree per millimeter. The Full Moon and the Sun both have angular diameters of about half a degree. You can see why ordinary snapshot images of the Moon or the setting Sun are so disappointingly small.

Table 4.1

Units of focal scale	Divide focal length (in millimeters) into
degrees per millimeter	57.3
minutes per millimeter	3438
seconds per millimeter	206 265

With a telephoto lens, which may have a focal length of 135, 250 or even 500 millimeters, and certainly with a telescope, it is usually more convenient to use focal scales expressed in minutes of arc per millimeter or seconds of arc per millimeter, and Table 4.1 shows how to calculate those values. Calculate the focal scale of the 200-inch mirror (660-inch focal length), and see if you get the right answer (see page 32).

Now look at Fig. 4.5 where we have added an eyepiece. To begin with, consider what happens to a parade of photons traveling (in parallel paths) from star A located practically at infinity and in a direction parallel to the central axis of the telescope. The front lens, or objective, again focuses the light at the focal plane whose distance from the lens is equal (by

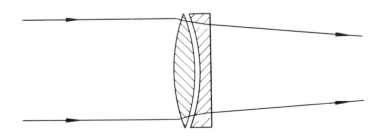

Fig. 4.4. Parallel rays of light of two colors will be focused at precisely the same point, as shown in this figure when a properly designed two-element lens is used. The focal point lies far to the right in this diagram.

definition) to the focal length of the lens. The eyepiece situated behind this plane gathers up the now diverging photons and alters their paths so that once again they are traveling parallel to one another. As you can immediately deduce from the figure, the distance from the focal plane to the eyepiece exactly equals the focal length of the eyepiece lens. Therefore the separation of the two lenses equals the sum of the two focal lengths. However, it would appear off-hand that the light has been de-magnified. It has not.

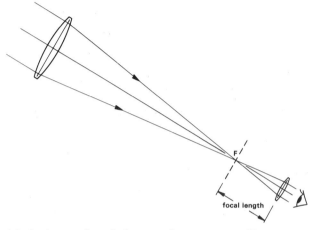

Fig. 4.5. In this simplest of telescopes, the eyepiece acts like a magnifier. For an observer with 20-20 vision, it is located a distance behind the focal point, F, equal to its own focal length.

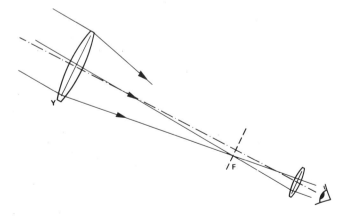

Fig. 4.6. When the simple telescope is pointed at a slight angle to the incoming rays of starlight, the focal point, F, falls off the optical axis (dash-dot line). Notice that not all the light will enter this particular eyepiece.

Let us consider what happens to light arriving from a star C, located off the optical axis. If we remember that (1) photons passing through the center of a lens go through undiverted, and (2) photons coming from star C also travel in parallel paths, we can make a second diagram, Fig. 4.6. The central photons pass through the center of the objective and, after a small displacement, go straight to the focal plane and on to the eyepiece where they are strongly refracted.

Next consider the path of those photons from Star C which end up passing through the center of the eyepiece. To have formed a good star image at the focal plane, this path must have passed through point F. Now we can continue the line of photons in both directions: to the right, straight through the center of the eyepiece, and to the left, until it hits the objective at point Y. The light from Star C entering the objective at Y had to be moving parallel with the central ray, and with that information we can complete the drawing of that particular path of photons.

In Fig 4.6, we have omitted (for the sake of clarity) the light paths from star A, but you can see that the angle between the rays from stars A and C which pass through the center of the eyepiece is larger than the angle between the rays from the same two stars as measured at the center or the objective. The ratio of these two angles is the *magnifying power*, and if you work through the geometry of Fig. 4.6, you should end up with the formula for magnifying power:

$$\text{M.P.} = \frac{\text{focal length of objective}}{\text{focal length of eyepiece}}$$

While there is no simple or inexpensive way to vary the focal length of the objective, having handy a set of eyepieces with different focal lengths makes it possible to change magnifications easily. Zoom eyepieces do exist, but they normally lack the flat, wide field of view provided by standard fixed-focus eyepieces.

There are limits on the amount of magnifying power your telescope can use efficiently, both maximum and minimum. The maximum magnifying power, up to about 200 times, is equal to around 15 times the diameter of the objective in inches. It is set by the *resolving power* of the telescope. This quantity tells you the separation of two just barely resolved stars of equal magnitude, and for modest-sized instruments depends directly on the size of the objective lens (or mirror). You can calculate it with this formula:

$$\text{resolving power} = \frac{4.5 \text{ seconds of arc}}{\text{diameter of objective (inches)}}$$

A 9-inch telescope can resolve two stars separated by a half second of arc – if the atmosphere is steady enough – and a magnification of 135 (9 times 15) is all the power one needs. In theory the Soviet 6-meter reflector can distinguish two stars separated by only one-fiftieth of an arc second, but the atmosphere would probably never permit it. A magnification of 200 times is usually adequate. Higher powers may occasionally make the viewing more convenient but no more distinct. The manufacturer who boasts a 9-inch telescope for sale with a magnifying power of 1000 times has to be considered a shady character.

It may be surprising to learn that there is a *minimum* recommended magnifying power below which one begins to lose light quickly. This is because the iris diaphragm in one's eye never opens to more than about 8 millimeters. If we look again at Fig. 4.5, we can see that as the eyepiece focal length increases, so does the width of the emerging bundle of light rays. If you do the geometry, you find that the minimum magnifying power is about three times the diameter of the objective in inches. Any magnifying power less than this results in an exit beam larger than the eye can accommodate. Some light never enters the eyeball, and faint objects appear fainter than they should.

Reflectors

As Newton first demonstrated, a mirror can also be used to focus light, and because all colors are reflected in exactly the same way it has the considerable advantage over a lens of being 100 per cent achromatic. All wavelengths focus at precisely the same point, at least for an on-axis star. The exact cross-sectional shape of the mirror is *parabolic*, which means it has a curve bending less and less rapidly as it gets farther from the point nearest the focus (see Fig. 4.7). Also, the extreme ends of a parabola must eventually end up (at infinity) parallel to one another.

A parabolic objective has at least three clear advantages over a lens system: (1) it is 100 per cent achromatic; (2) the purity of the glass need not be high since the light never actually passes through it (up until the end of the last century, most telescope mirrors were made of metal); and (3) only one face needs to be finished optically, the aluminized front surface.

These advantages result in considerable savings of money, particularly for large telescopes. Furthermore, one can support a mirror from underneath, thereby making it possible to make very large objectives. Big unsupported lenses will sag and can even break under their own weight. The achromatic objective of the world's largest refractor, the 40-inch at the Yerkes Observatory

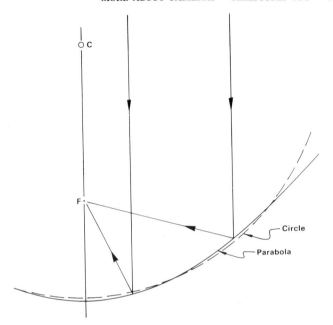

Fig. 4.7. A parabolic mirror focuses all light rays traveling parallel to the axis of the parabola at one point, no matter what the color. This figure shows the shape of a parabola compared to a circle whose center is at C.

near Chicago, sags nearly an inch at the center when the tube is pointing vertically. The basic limitation on the size of objective mirrors is the problem of transporting the finished product to the site. If telescopes with single mirrors much larger than the Soviet 6-meter reflector are to be attempted, it will be necessary to build the glass foundry right next to the dome. Hence the current interest in multi-mirror systems.

For all but the largest reflectors there still exists one problem, namely getting the light, which is reflected back from the primary mirror, out of the tube so that no serious obstruction to the incoming light will occur. Newton's original solution (see Fig. 4.8) remains the simplest and least costly: intercept the converging light rays with a flat mirror set at a 45-degree angle and reflect the light out of the tube at the top end. The flat surface of this secondary mirror alters neither the focal length nor the focal ratio; it only redirects the light. Approximately 15 per cent of the light is blocked by the secondary of the *Newtonian* reflector, but no serious degradation of the final image occurs.

Another solution to the problem was devised by the nineteenth-century French optician Louis Cassegrain, who cut a central hole in the primary mirror and redirected the converging

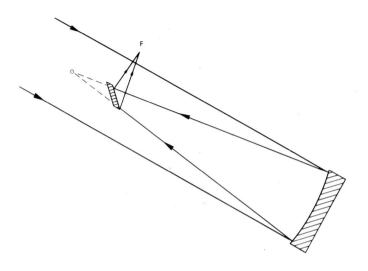

Fig. 4.8. The Newtonian reflector uses a flat mirror to divert the light out of the telescope tube. Except for blocking out approximately 15 per cent of the incoming light, this secondary mirror does not appreciably affect the quality of the image falling at F.

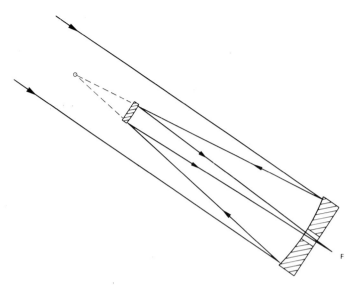

Fig. 4.9. The convex shape of the secondary mirror in this Cassegrain telescope extends the focal length and increases the effective focal ratio of the telescope. A Cassegrain is well suited to observing the surface features of the Moon and planets.

rays back down through this hole. The surface of the *Cassegrain* secondary obviously should be convex (see Fig. 4.9) to slow the convergence rate of the photons coming from the primary and to re-focus the light beyond the back side of the main mirror. The correct cross-sectional shape for this secondary is a *hyperbola*, a curve where the ending lines never converge nor become parallel.

The *Cassegrain* reflector has two small advantages over the Newtonian: (1) it provides a more convenient location for the focal plane, and (2) it can have a longer focal length. Actually, by having several spare secondary mirrors it is possible to have available several focal lengths for one telescope. Moreover, by clever positioning of one or more flat mirrors, it is possible to have the focal point remain stationary no matter where the telescope points, a particularly useful characteristic if heavy or bulky equipment is used in conjunction with the telescope.

Perhaps the biggest disadvantage of a reflector is that the field of good, sharp focus is small. For the 200-inch Palomar it covers an area about 30 millimeters across. With a focal scale of 12.3 seconds of arc per millimeter, this amounts to just over 6 minutes of arc. However, opticians have found lens systems that go just in front of the focus and which greatly enlarge the size of the field. On most of the larger reflectors, such correctors are available.

Bernard Schmidt, a one-armed telescope maker from Ger-

many, reasoned that because a spherical primary mirror had no uniquely definable axis like a parabola or hyperbola, one could in principle get a larger field, if only the proper correction lens could be devised. In 1930, he found that such a lens could indeed be made (Fig. 4.10) and, as a result, *Schmidt* cameras or telescopes are able to take beautiful wide-angle photographs. While the field of good definition for a fast Newtonian is only a few minutes of arc, a standard Schmidt can obtain good star images over a 5- to 10-degree field, and specially altered systems are able to go to 50 or 60 degrees. It has been said that besides having only one arm, Schmidt was also fond of the local brew. Knowing well that spherical surfaces were far easier to make than parabolic ones, he was actually hoping to find a way to avoid long hours of hard work. His spectacular discovery was, in a sense, unfortunate because Schmidt camera optics are far more time-consuming to make than a single paraboloid mirror.

In 1944 the Russian optician D.D. Maksutov described yet another way of correcting a spherical mirror. His solution was to put a thick lens with little power but steeply curved surfaces at the upper end of the tube (Fig. 4.11). This type of corrector, called a *meniscus* lens, is spherical on both sides and therefore relatively easy to make. However, because of its thickness, it requires high-quality glass. It is a typical 'catadioptric' system.

Various combinations of these optical systems are possible.

Fig. 4.10. The correcting lens or plate of a Schmidt camera is usually smaller in diameter than the primary mirror. Thus the parallel rays of starlight entering off-axis in this diagram all strike the mirror and are focussed at F. For best results the photographic film or plate must be curved as shown here.

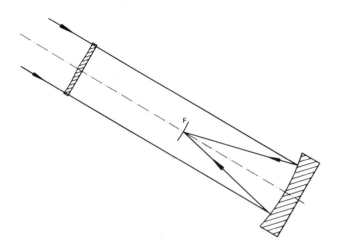

Fig. 4.11. The Maksutov telescope, like the Schmidt, has a wide field of good definition. The secondary mirror is often an intergral part of the meniscus lens.

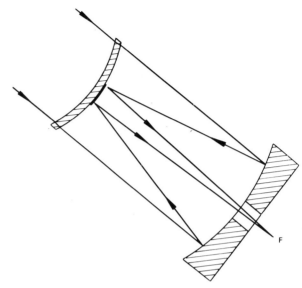

Fig. 4.12. The Schmidt–Cassegrain conserves the advantages of both types of optical systems. This drawing, adapted from a manufacturer's design, shows also how the weight of the mirror can be appreciably reduced.

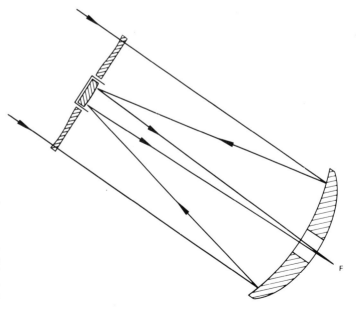

Fig. 4.12 shows one configuration, the Schmidt–Cassegrain, currently one of the most popular designs on the market.

Numerous other types of reflecting systems exist and undoubtedly more will be developed in the years ahead. In future chapters we will describe how telescopes can be conveniently mounted and motor-driven to follow the stars, what it actually is like to use some of the larger telescopes, and finally how one goes about grinding and polishing one's own telescope mirror. *Reader be warned*: There are many who have found telescope-making addictive. The satisfaction of having made with one's own hands a mirror whose surface is accurately shaped to within a few millionths of an inch, and which can be used to view the rings of Saturn, the satellites of Jupiter, the arms of spiral galaxies, and countless stars beyond the reach of the unaided eye, must be experienced to be believed.

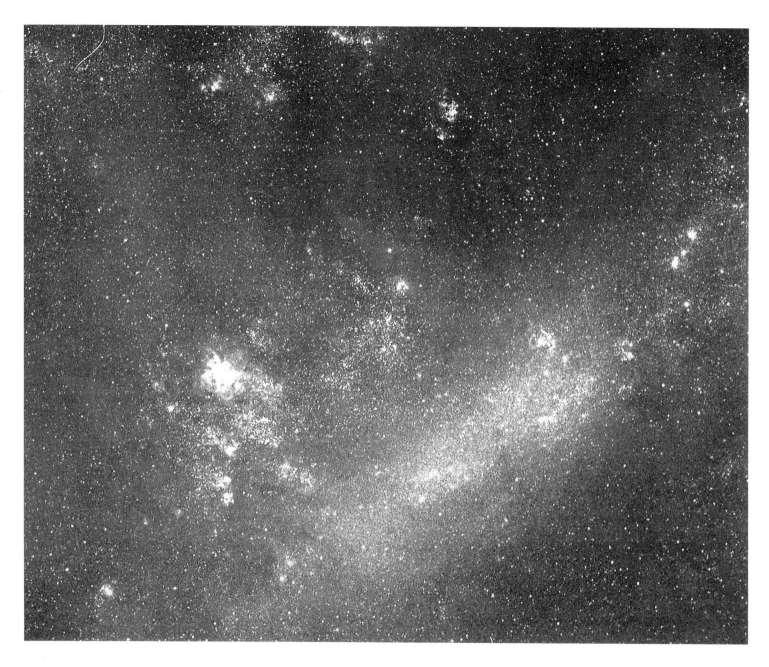

Fig. 5.1. The Large Magellanic Cloud, our nearest companion galaxy. This photograph was taken with a small Schmidt camera by Manuel Lopez Alvarez in Buenos Aires.

Chapter 5: Catching starlight

You can be quite sure that at this very moment someone, somewhere, is competing in the grand prix race of film speeds. 'Faster and faster' is the slogan. In 1915 no one at the Indianapolis track could have imagined the performance of a sophisticated, rear-mounted powerplant when Ralph de Palma tooled around the course at a hair-raising, record-breaking 89.84 miles per hour. And my great-grandfather, sitting rigid and unsmiling for 5 seconds posing for a now-yellowed photograph, could not have dreamed that his future offspring would take instant color photographs in one-five hundredth of a second. Let us not stop there, when we already know that great-great-great-grandchildren will be able to record moving color images on video tape for immediate playback, amplification and storage.

Time passes quickly; but try to tell this to an astronomer shivering on a mountaintop at the guide-scope of a large astrocamera, while making a one-hour exposure and he's likely to say that time is dragging and that his feet are cold. Because scientists need ever greater camera and film emulsion speeds, the amateur community also benefits.

It is tempting to bypass black-and-white photography altogether and to take the giant leap to color right away. Television made the change from black and white to color in a surprisingly short time; early monochrome sets are joining Kodak Brownies on collectors' shelves.

For astrophotography, however, black-and-white film – 'something or nothing records' – will be with us for a long time to come. That is why this chapter covers it, as well as color photography. Color plays a vital role in the study of the Universe, especially when prisms are used to break up the light from distant objects into rainbow strips of their 'spectra'. Analysis of these colorful bands allows us to detect the stuff of which stars and comets – even meteors – are made. Still, the photographing of faint objects at ever greater distances will long remain an important goal. If observers at the Palomar Observatory can see the light of a candle at 20 000 miles, they will be content – for a little while longer – to do so in black and white. After all, what are 10 or 20 Earth-years in terms of astronomical time, when we consider the marvels of ultra high-speed color recordings, which are sure to be part of our future?

There are many different types of film both in the black-and-white category and in the color group. They have one thing in common: they are all rated on scales for their 'speed' or sensitivity to light. This rating is printed on the film casettes and boxes in the form of a number, usually preceded by the letters ASA, ISO or DIN. These stand for 'American Standard Association', 'International Standards Organization' and (translated) 'German Industrial Norm', respectively. This establishes 'how fast is fast'. Let's use the ISO scale to explain that the lower the ISO number, the less sensitive (or slower) the film emulsion; the higher the number, the more sensitive (or faster) the film. A film with an ISO rating of 25 is much slower than a film rated at ISO 800.

It's easy to remember:

LOW ISO NUMBER = LOW SENSITIVITY.
HIGH ISO NUMBER = HIGH SENSITIVITY.

Light sensitivity indicates how long it takes for the emulsion with which the clear film is coated to record the light which falls on it. One would think that since the stars, clusters and nebulae pretty well stay in one place for long times it would be easy to photograph them if one held the camera steadily pointed at some point in the sky, and left the shutter open long enough. This assumption is surprisingly correct today with the possible combination of fast films and camera systems. It is for this reason that we will address color photography first. In all likelihood it is color film which is in most cameras today. One thing must be stressed here: film with which one takes family scenes is not quite right for astrophotography, but it will work for a start if you want to shoot out a roll on astronomical subjects. Do yourself one favor though, start a little booklet right now, and record in it what type of film was in your camera, and what type of settings the light meter suggested you use for that glorious sunset scene which you will have no trouble recording. One more suggestion, if the light meter tells you to shoot at f/11 with an exposure of one-hundredth of a second, please 'bracket' this exposure with two others. Take one at f/16 and another at f/8, each with an exposure time of one-hundredth of a second. Whatever you do be sure you write down the order in which the three exposures were taken. When the processed film comes back you will instantly find out the function of the f adjustment. Remember, it stands for 'fastness'. What you will have done by turning the little ring on the lens, or by changing the controls in some other new-fangled way, will instantly emerge: you will have made your lens 'faster' at the f/8 setting and 'slower' at the f/16. If the 'iris' is visible by looking in your lens from the front you will see that the size of the hole through which the light must pass is larger at f/8 and smaller at f/16. If you cannot see the 'iris', which surrounds the 'aperture' (opening), from the front end, you can usually observe the different settings from inside the camera while it is open for reloading. The pictures will immediately show how the larger aperture (f/8) allowed more light to pass through the lens than did the smaller (f/16). They will also suggest slight modifications over the settings which the light meter dictated.

WARNING: NEVER POINT YOUR CAMERA AT THE SUN EXCEPT WHEN IT IS ONE SUN-DIAMETER ABOVE THE HORIZON OR LESS; AVOID LOOKING AT THE SUN DIRECTLY OR THROUGH A CAMERA IF IT IS HIGHER IN THE SKY. DAMAGE TO YOUR EYES AND TO YOUR CAMERA COULD OTHERWISE OCCUR.

If you still have pictures left in your camera, how about trying to shoot a *thin* crescent Moon? Now the light meter becomes useless, which is why it is best to set the lens at its largest opening. Leave it at this setting and 'bracket' the exposure times. To do this you will need two things: a tripod to hold the camera steady for the longer exposures, and a cable-release to hold the camera open for longer periods of time while setting it on 'B'. You will notice that the largest possible opening of your lens, which may range from f/1.4 to f/2.8 depending on the make of the camera, is in effect the rated speed of your lens. This is the f setting which you want to use not only for your lunar photography but for all astrophotography hereafter. In this fashion the only variables remaining are exposure times and the speed of the film.

For lunar photography use the 'longest' lens you have. Mount the camera on a tripod or secure it firmly with a clamp after pointing it at the Moon wherever it may be, low or high in the sky. But don't place the Moon dead-center in your viewfinder, put it off to one side, because you might catch some distant horizon along with the lunar disc or, quite unexpectedly, some star images which will help compare film speeds when you go to faster emulsions.

Now connect the cable-release and start by trying a half-second and a one-second exposure. (These are built-in settings in many cameras. Otherwise, set the camera on 'B' and count it out: 'one and one' for one second.) Then advance the film and shoot another exposure of 2 seconds: 'one and one, one and two'. Advance again and now try an exposure of 4 seconds: 'one and one, one and two, one and three, one and four'. Go ahead, shoot the works, try an 8-second exposure, one of 15, and one of 30 seconds. You will have doubled the length of each exposure over the previous one. Immediately write down all that you did in your little booklet: film-speed and type, f/stop (wide open), length of exposures, and in what order the exposures were taken. You might add what phase the Moon was in (make a little sketch) and some notes on weather or humidity. Every entry will become invaluable later and save you time and money. You will learn more about photography when you review the processed pictures or slides than any manual can teach. The Moon will be small in such picture-taking.

The message which your notebook will make clear has to do with the amount of light which passed through your lens. Since we will always be working with the maximum possible aperture (largest lens opening and fastest possible lens) when we take astrophotographs, the exposure time becomes important as any stars in the picture will show.

There are limits to how long one can leave open the shutter for an exposure. The first problem has to do with light pollution, especially near cities. The great amount of 'ambient' light, which is the light given off by street lamps, neon signs and parking-lot illumination, and which just seems to hang in the air, is the main culprit. If the lens is left open too long, this light will gradually accumulate on the film and expose it.

Another reason one cannot leave a camera to expose indefinitely, even under favorable dark skies, is that the film gets 'tired' after a while. It develops 'film fatigue'. In general, the faster the film the sooner this fatigue sets in. Professionals refer to it as 'reciprocity failure', an impressive term when casually dropped at a cocktail party. Such words will greatly impress the uninitiated. Do not let them intimidate you. One can best explain 'reciprocity failure' by comparing the high-speed film to a person trained to run the mile in one minute. The runner will achieve the speed needed to perform this extraordinary feat but will not be able to continue at anywhere near this rate.

When we use standard off-the-shelf films for astrophotography, we are asking the emulsion to run a marathon at minute-mile speeds. It is for this reason that the film will collect most of its starlight in the first minute or so. It will continue to store additional light in the succeeding minutes, but at a much slower and decreasing rate.

A final reason why it is not possible to take star photographs of indefinite length is that the stars seem to move (as noted in Chapter 1). Today we know it is the Earth which turns. Photography makes it appear as if the stars moved in the heavens (see Fig. 1.1). Because of this, star images will 'trail' on longer exposure photographs taken with standard cameras. Many interesting and useful star-trail photos can be obtained in this manner with exposures of long duration. One-, 2- or even 3-hour exposures are possible.

You may well want to try to take such a star-trail photograph, for which a tripod is desirable but not even necessary. Just lay the camera on a table in a dark area, point the lens skyward and open the shutter with the kind of cable release which permits you to 'lock' the shutter in the 'open' position. Be careful that no one shines a flashlight near the open camera and that it remains untouched for the entire duration of the exposure. After any length of time of your choice – from minutes up to half an hour – close the camera and again note down exactly what you did. The resulting photograph will yield invaluable information concerning the motion of our planet and exposure times for the type of film which was used.

In color films the principal difference between the slow-to-medium types and the highest speed variety may have to do with color fidelity. This is not critical. Our first concern should remain the maximum speed of the film. Let's think about the choices offered in color.

As a rule of thumb, slide-film seems the ideal way to go for astrophotography. By using a projector in a darkened room one can throw very large images on a screen to create a window to the sky with planetarium fidelity. It is always possible to make prints from transparencies, but things get a little more complicated when one begins with negative color film which is geared to the making of color prints only. The ISO 800, 1000 and 1600 speeds are ideally suited and most manufacturers of photographic materials offer these in rolls of 20 or 36. It makes little difference which one you pick, some people swear by Fujichrome, others are Kodak Ektachrome *afficionados* or favor other products. Perhaps you want to start by trying a roll of 20 exposures of each and then staying with the one which pleases your color sense. It is best to become acquainted with a film and then to stick with it. Your notes will tell you how it performed in past situations and permit predictions about future exposures.

You may wish to start exploring the skies with color film, but eventually you may want to consider the black-and-white alternative which opens up exciting new possibilities. Inevitably this should involve film processing which includes not only easy film development, but takes you into the wonderful world of making your own enlargements.

Black-and-white films are much less costly and printing papers for enlargements are also relatively inexpensive. Bear in mind that very few commercial photolabs today are equipped to do anything more ambitious with black-and-white material than to develop film and then make standard-size prints. You will want 8 inches × 10 inches enlargements which alone allow meaningful images with which you can study what you photographed or build a collection of the constellations. Such blow-ups can give you images which look as if they had been taken through telescopes.

Enlargements from your astronegatives would be called 'custom work' in a camera store, and could easily lead to difficulties. For one thing, the laboratory technician may not realize that the picture which you took which has the meteor (shooting star) in it, is very special. They may print or enlarge it without understanding or feeling, – or both. By the second time, you will have tried to explain to the clerk behind the camera-store counter that those seven stars off to one side of the enlargement are the 'Pleiades' in the constellation of Taurus the bull, that this grouping of stars resembling a tiny dipper is 430 light years away and – above all – that this prime object should be in the middle of the picture, with the shooting star off to the side. You will probably be told 'Look buster, if you're so smart, why don't you do it yourself?'

Do it yourself, because black-and-white processing is really easy. You can quickly become expert in the field and the items you need to purchase to 'do it yourself' will pay for themselves over and over again. The heart of the whole set-up is an enlarger, which you must have. The rest is trays, tanks and tongs and some chemicals with simple instructions. Don't forget: do not buy a telescope yet. Buy an enlarger instead. You can magnify the sky which you recorded with your standard camera, and view it clearly on a rainy Sunday afternoon.

Get the simplest (and shortest) booklet on processing black-and-white film, which tells you how to set up a darkroom in your bathroom or kitchen, and follow the simple instructions. You will be delighted to find that you can even make professional-looking black-and-white prints from your color transparencies.

An alternative possibility presents itself for those who live in larger cities: there are photolabs or studios which charge a per-hour fee for the rental of darkroom space. These facilities are usually set up for use by a dozen customers at once. They will even develop the film for you and usually give you a hand getting started. Normally they also suggest that you buy your printing paper from them. This is not only fair, but a good idea because they have large supplies of fresh material.

I have found that in 15 minutes someone who knows what they are talking about will show you how to put your negative in the enlarger, focus it, make a test strip, and how to develop, fix and wash the print. After 30 minutes, your very first, very own enlargement of the Pleiades, just as you wanted it, will be floating glossy and wet in the fixing bath. You'll quickly make three more copies to share or give away, while the negative is still focused in the enlarger.

You may think that all darkrooms are totally black. Nothing could be farther from the truth. Many are lit with deep red romantic safety lights which do not affect the photographic papers but allow you easily to distinguish between the long-haired co-ed, who takes pictures at Trade-Tech, and the longer haired commercial artist, who finds it convenient to use the rented facilities.

Try it, you will like it. You *will* like it. You will like it so much that, when after a short and predictable while you start shopping for your own processing equipment, you will know exactly what you want and need, and where to get it. You may even look into color processing equipment and check out an enlarger which will allow you to do your own color work.

Do not hesitate to put your Polaroid and Kodak 'instant cameras' to astronomical use. In addition to color, there are extremely fast black-and-white films available, even some where in addition to the instant print, you get a negative which you can use as a permanent record and for enlargements. New and better films are being offered constantly.

As a general rule, in photographic matters you can be sure that any reputable camera store (go to the biggest one in your town) will have someone there who can help you and will enjoy demonstrating – freely – that there is really very little in photography that you cannot master. Ironically such a person may not be acquainted with astrophotography where you can become expert as soon as you realize that none of the complexities of regular daylight photography need to be studied to get you started. Nevertheless, get a simple book on the subject. Usually the manual that came with your camera contains all you need to know in the beginning.

One final word: the notebook in which you will keep records of everything you do will be your most important and precious manual. Experience is your best teacher and the only real guide to better results.

Fig. 5.2. Star cloud, M24, in the Milky Way.

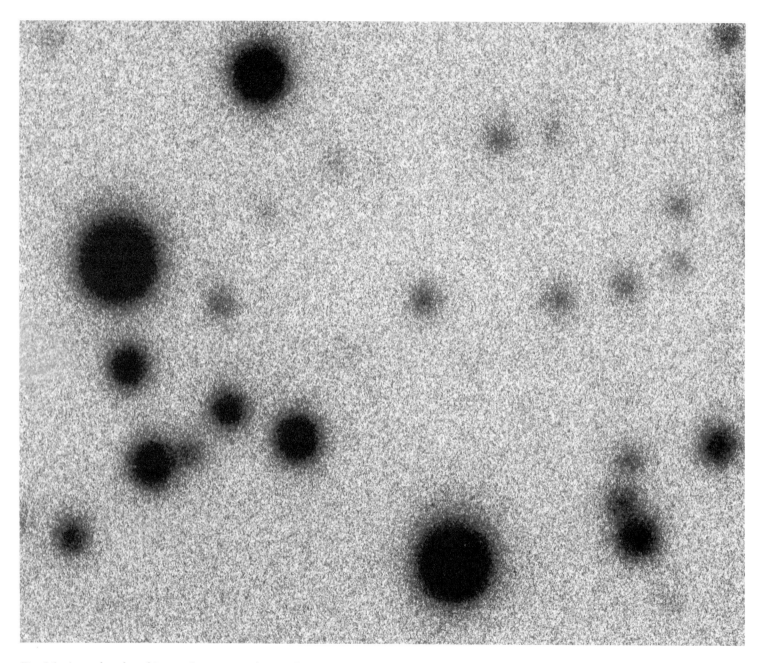

Fig. 6.1. A greatly enlarged image of a star on a photographic negative shows it to be composed of dozens or even hundreds of emulsion 'grains', each averaging 0.01–0.02 millimeters in diameter.

About film

CHAPTER 6

When you next buy a roll of Tri-X or Fujichrome 400, think for a moment about the problems you would encounter if the fastest film in the store had an ISO (or ASA) rating of only 0.0001. Louis Daguerre had those problems in the 1830s when he invented the daguerrotype. So did Bond and Whipple at the Harvard Observatory in 1850 when for the first time they photographed a star. Furthermore, they had to do *everything*, first 'fuming' a carefully polished silver plate with iodine to make it light-sensitive, then developing it by fuming it with mercury vapor, a potentially lethal procedure which drastically shortened the lives of many an early photographer.

The most important constituent of the yellow plastic-like substance (the emulsion) that coats modern black-and-white film is actually silver; it is the ingredient which turns out black in the developed negative. In unexposed film, the silver starts out as part of a yellowish molecule, a salt, like silver chloride (table salt is sodium chloride) or silver bromide, suspended in a hardened gelatin that adheres to the transparent plastic film or to the printing paper. Silver salts are photosensitive which means that under the action of light the chemical bond holding the two atoms together is weakened. 'Developing' is the process by which the weakened bond is finally broken and the chlorine (or now usually bromine) is washed away. When you 'fix' a plate, you dissolve away the unexposed silver salt leaving clear areas of gelatin where no light struck and blackened silver everywhere else.

Developing your own film

Setting up a home processing laboratory for black-and-white film requires no special talents or expensive apparatus. If you have already experienced the satisfaction of developing your own films exactly in the way you want, then you can skip the rest of this section which is nothing more than a quick course in film processing. If not, then read on, or perhaps take a course from a local expert and learn some of their special tricks and time-saving techniques as Chapter 5 suggested. You have nothing to lose but the scratches and streaks which your downtown commercial film processor too often provides.

To begin with, buy a standard developer like D-76, or Rodinal, some 'fixer', and a film tank, one of those light-tight metal or plastic cylinders with a close-fitting lid and a special spool inside on which film is loaded. Nothing more is needed. After developer and fixer are mixed according to the straightforward labelled instructions, the step-by-step procedure that one follows goes like this:

1. First find a room which you can make completely dark. Either lock yourself inside or else make absolutely sure that everybody knows that you are not to be disturbed until told otherwise. The first time that you are in your darkroom, let your eyes become fully dark-adapted (at least 10 minutes), and then seal up all light leaks under the door or around windows. If you have a safe-light, do *not* use it when handling astronomical film. Safe-lights are for paper prints only.

2. While in the dark, load the film into the developing tank. The first ten times are the hardest, but by practising initially in the light with some already developed (or spoiled) film, you will soon learn the tricks. You should always use an absolutely clean and dry wind-up spool – and trim off the corners of the film before loading.

3. After making absolutely certain that the tank top has been screwed on properly, turn on the lights. From here on everything can be done in the open.

4. Pour in the developer through the special opening in the tank top and start gently but continuously agitating the contents either by twirling the spool or by rocking the tank back and forth. Continue for the number of minutes specified on the developer instructions.

5. Pour out the developer and run fresh, clear water into the tank for about 30 seconds. (Some prefer to use a 'stop bath' after the developer, but it is not necessary.) The developer may be saved and re-used for several rolls of film if stored, tightly capped, in a dark or opaque container, like a beer bottle, and used within a month or so. The water temperature should be within a few

Fig. 6.2. (overleaf, pp. 42–43) Hypersensitizing a photographic emulsion sometimes can perform miracles. At least it can seem that way on a cold and windy night. These two exposures of the southern Milky Way were made in exactly the same way (Kodak 2415 film, f/1.4 55-millimeter focal length, 3-minute exposure), except that one film had been hypersensitized in warm forming gas for a week, increasing its speed six times.

degrees of both the developer and the fixer that follows, preferably 20°C.

6. Pour out the water and pour in the fixer or 'hypo'. After about a minute of agitation, you may open up the tank and watch the process. Continue fixing for at least twice as long as it takes to clear the film of its milky appearance (a total of 3 to 10 minutes).

7. Pour the fixer back into the container and rinse the film in water for about one minute. Repeat at least five times. Continuously running water can be used but is risky if hot water is being added to bring temperature up to that of the developer and hypo (as it should be). A flushed toilet somewhere else in the house could ruin your night's efforts. The fixer keeps well; discard it when it takes more than 7 or 8 minutes to clear the film.

8. Finally rinse the film with a very weak solution of detergent, preferably the 'non-spotting' dishwasher variety. Photo companies market a liquid especially for the purpose.

9. Remove the film from the spool and hang up to dry in a non-dusty place out of the traffic pattern of the house.

10. Admire your results – or analyze what went wrong.

Two precautionary notes:

1. Handle wet film with extreme care, since in that stage the emulsion is only slightly less fragile than warm butter.

2. Do not mix chemicals in your darkroom. Fine dust from the developer or hypo powders can badly pock-mark negatives.

As Chapter 5 pointed out, the heart of a darkroom is the enlarger which will enable you to show off your first pictures to their full advantage. With it you need trays (three are enough), a red safelight, and of course printing paper. The fixer and developer remain the same, as do the basic procedures, only now you can vary exposure and developing times and even kinds of paper to achieve the desired results. Experimentation is the key word; the only additional advice we give is an obvious one: the vital component of the enlarger is its lens. If you have a little extra to spend, invest it in the enlarger lens. A cheap lens can produce the strangest looking star images, especially in the corners of the print.

Color film

In the mid-1930s Eastman Kodak began marketing Kodachrome film which contained three emulsions sensitive to blue, yellow and red light and separated by dyed layers. In the process of development, the blackened silver is bleached out as before, but now left behind are the dyes which remain to provide the proper color combinations on the final slides. 'Kodacolor' and other emulsions of its type (most have names with the suffix '-color') are designed to provide a color negative from which positive prints can be made with relative ease. We would advise you to stay away from processing color negative materials unless you plan to get involved with the fascinating but much more complicated techniques of color printing, although many kits are now available.

For most of us then, the main disadvantage of color film is that others, possibly a machine, will do the developing. It usually takes one or two days to get your film back and you have little control over how the film was processed. If the time element is especially important, you might want to experiment with self-processing film. Just follow the instructions blindly. The time you gain may net you a discovery.

Emulsion characteristics

Color sensitivity

Nearly all the black-and-white films marketed nowadays are panchromatic – sensitive to all colors from ultraviolet to red. However, one can still find black-and-white emulsions with limited color response, especially on films (and glass plates) marketed for scientific or astronomical use. In fact, to determine the color of a star or stars using black-and-white materials, the professional astronomer will make exposures on two or more different types of emulsion, because obviously a red star will register more strongly if the film is most sensitive to red light, while a blue star will show up best on blue-sensitive film. Actually, a 'red-sensitive' emulsion is not only sensitive to red light; it also responds to orange, yellow, green, blue, violet and ultraviolet light, and thus is *panchromatic*. If you want to see only red with your camera, you have to use a red filter with 'pan' film. Similarly, a yellow-sensitive emulsion will record yellow, green, blue, violet and ultraviolet light; a blue-sensitive film is sensitive to blue, violet and ultraviolet light. More on this later when we discuss films (and plates) made especially for astrophotography.

Film contrast

The contrasty black-and-white images we associate with Charlie Chaplin movies and with stills taken early in this century got that way because the films used were slow. The degree of contrast, especially important if you are photographing an extended object like the Moon or the Orion Nebula, can be altered considerably by the developer, but at the same time the film speed is changed too. If you shorten the development, the picture comes out looking flatter, lacking contrast.

The better film stores sell special black-and-white emulsions such as Kodak High Contrast Copy (ASA 64) or Type 2415 Technical Pan. Their relatively slow speeds, unless hypersensitized (see below), make them useful mainly for the brighter objects in the sky. Try some shots of the Moon soon; the results will be startling. Or use such films with a very fast camera lens – f/1.4 or faster – and shoot the Milky Way.

The above words about contrast apply to color emulsions as well, but because the aim of the manufacturer and the film processor is to produce lifelike pictures, the available range of contrast is small.

Color balance

One property of color film is that after a long exposure taken of the glorious star-filled night sky, the background may show up in some unlikely shade of turquoise or purple. The problem is that the manufacturer adjusts the sensitivity of each of the three-layered emulsions to be correct for an exposure of a fraction of a second. Expose for a half hour and you take your chances: maybe the color balance of sensitivity will be correct; maybe it won't. But this is not important. What is important is that a comet may have been arrested in its motion for all time.

Granularity

On entering a photography store, the novice star-photographer will quickly rivet his or her attention on the fastest films available. ISO values of 10000 and up are now on the market. But beware! Fast film usually is grainy; original slides and prints will look like portions of an enlargement blown up too many times. Under magnification, star images show up like fuzzy spots rather than the pin pricks produced by fine-grain emulsions. So here is another trade-off: fast film and graininess, or slow film and fine grain? Because the triple-layer emulsions of color films smooth out the granularity noticeably, you should shoot in color if you want to use the highest film speeds available.

Astronomical emulsions

Manufacturers provide black-and-white emulsions which are specially intended for astronomical work. One general property of ordinary films which conflicts with the needs of the astrophotographer is that their speed decreases as exposure time increases. Chapter 5 described this trade-off in emulsions in terms of fatigue – like a sprinter running the mile. Tri-X may have an ISO rating of 400 when you're shooting at a fiftieth of a second, but it drops to 100 if your exposure time is lengthened to 10 minutes, and it's down to 50 after an hour's exposure.

For the long-suffering astrophotographer, Kodak provides a number of emulsions, some of which are available on film. (All are available on glass plates, the choice of most professional astronomers because of their dimensional stability.) Such special films and plates can usually be identified by the small letter 'a' appearing in their type number. These exotic emulsions can be purchased either in bulk quantity directly from the manufacturer, or packaged in the familiar 35-millimeter film casettes from a few specialty companies who advertise in magazines like *Sky and Telescope* or *Astronomy*.

Hypersensitizing

It has long been known that the performance of an emulsion depends to a certain degree on temperature, humidity and other factors. Early attempts at increasing the speed of a film included pre-exposing it so that the entire film had a barely noticeable fog on it to begin with. The idea was that light from a faint star would not have to work as hard to blacken the emulsion. This method works, but others are better.

Considerable improvements can be achieved by keeping the emulsion very cold during the exposure, usually at the temperature of dry ice. This procedure both increases the speed, especially at long exposures (it inhibits reciprocity failure) and, with color film, it improves the color balance. The main problem with this technique is that the moisture in the air around us rapidly condenses on anything at the temperature of dry ice – and immediately freezes. Elaborate precautions must be taken to eliminate this problem. Several companies manufacture 'cold cameras', film- or plate-holders which use a heated window to keep away the frost and dew. You might build your own.

Baking emulsions for a number of hours at a very gentle temperature (around 60°C) also increases speeds and reduces reciprocity failure if the film is used within a day or two. Even better results are achieved when the oven chamber is filled with nitrogen gas or 'forming' gas, a safe mixture of nitrogen and hydrogen (*never* use pure hydrogen). You might want to experiment with various procedures and develop new recipes.

A simple effective technique which requires neither baking nor chilling is to 'soak' the film or plate in an atmosphere of pure nitrogen or forming gas for several days or even weeks. Improved hypersensitization results if the soaking chamber can first be evacuated for a few hours and the gas allowed to flow through the chamber. The basic idea is to remove all the moisture in the emulsion and to replace it with molecules of a gas which do not have the desensitization properties of water vapor. Nitrogen is one of the best of these.

Most importantly, go take some pictures of the night sky *now*. Worry about the details later, for what's up in the sky tonight may not be there next week.

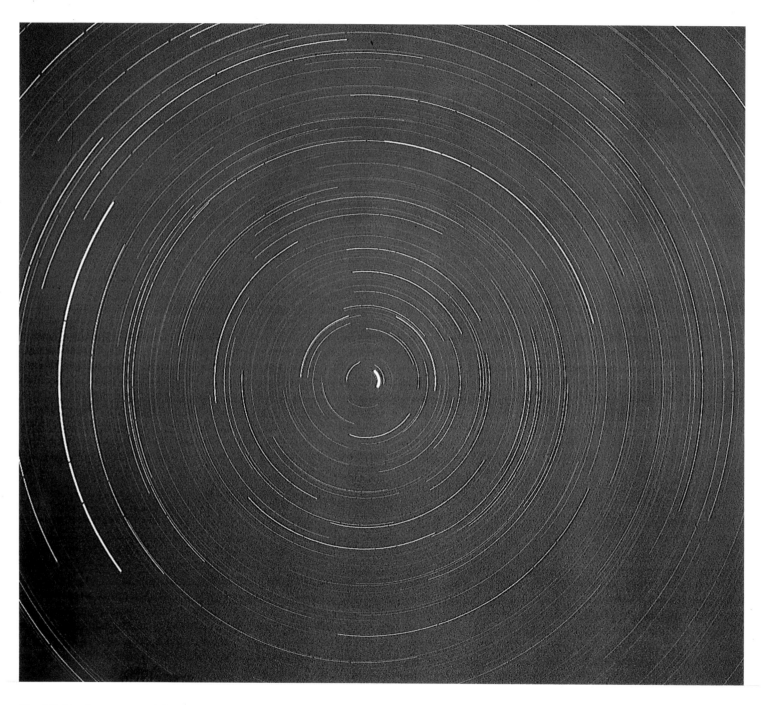

Fig. 7.1. North star region, 3-hour exposure. 40-millimeter lens, f/4. High-speed Ektachrome slide-film, ISO 400.

CHAPTER 7

A picture a night, a film a month

Load your camera with the fastest film available. Find a piece of chalk and write on any surface FILM No. 1, the date, and the type of film you're using. Add your name, address and phone number to be on the safe side, just in case the film gets lost in processing. Shoot this information twice with normal daylight settings some time before sunset.

The reason for this procedure is simple. First of all, the data on these first two frames will become an integral part of your permanent records. Also, in the case of color slide-film, these two frames will serve a very important additional purpose; when the processing lab puts its film into the 'mounting' machine (the device that puts each slide into a neat, cardboard frame), the operator may not be able to tell where your film begins if your first shot is a night exposure. He will then guess where the first frame is and you may get back transparencies which straddle two frames. The indexing method described will save you – and the processing technician – much grief.

Next, set the lens to the widest possible opening (you can check this during loading by looking into the lens while turning the aperture ring), set the shutter speed dial on 'B' and turn the focusing ring to infinity, ∞. You may want to secure these lens settings during night photography by taping them down with some electrical tape. This will prevent accidental errors from being made.

If you have any doubts about your camera, refer to the instruction manual that came with it. This booklet is important because it yields much valuable information which will be useful to you as you proceed. If you have lost the brochure, write to the manufacturer; they usually can help by sending you a replacement or a substitute for very little money. Even if you can only get a later edition, don't forget that the basic system of your camera has remained unchanged and that only the fancy appointments, which you do not need for your work, keep changing.

Step out into the night – yes, any clear night will do just fine – and select your heavenly target. There are three kinds of good nights: the bright ones when the moon is full and round or nearly so – we can view it through binoculars or photograph it through simplest telescopes; the crescent or Half Moon nights which are a little darker; and the New Moon nights when Luna, which revolves around Earth about once a month, lies between us and the Sun. In time you will begin liking these darkest nights best because they give you the optimum conditions during which to observe or photograph stars or other distant worlds. A black velvet background will make even tiny stars glow like jewels. It is almost always possible, within the span of any one night (except during Full Moon), to find a little time when there is complete darkness. It may be conveniently after dinner before the Moon rises, or past midnight, after the Moon has set. It depends on the lunar phases. Modern astronomers have said that we yet talk about the Sun and Moon 'rising' and 'setting' just as if we were still in the Dark Ages, back when the world was considered to be flat and to have 'four corners'.

In astronomy there is no 'up and down', only 'in and out and around'. Those are terms that pilots use today and space travelers will employ tomorrow. We must remember that the Sun only *seems* to rise or *appears* to set. What actually happens is that the Earth, by turning on its axis, brings the Sun into view once every day, which has led to our measuring daily time 'by the Sun'. It also gives us a cycle of daytime and night, measured in 24 hours.

As the Earth turns once every 24 hours, it moves about the Sun approximately once every 365 days. That is the motion on which our present calendar is based. All the while – to add to the mystery and the magic – the Moon revolves around the Earth, while the planets, like Earth itself, keep circling the Sun. The rate of travel for each celestial body is different but a majestic order governs them all.

For our immediate purpose, only one motion need concern us. It is the rotation of our Earth on its axis which makes the heavens seem to sweep from the eastern to the western horizon. Let us take our first photograph to demonstrate the concept as shown in Fig. 7.1.

The axis on which our planet turns – that imaginary line which connects the South Pole to the North Pole – continues on out, pointing at an abstract point in space called the North Celestial Pole (NCP). There is a star near there called Polaris, which is really part of the constellation of Ursa Minor (the 'Little Bear') and is the forty-sixth brightest star in the heavens. There are even two

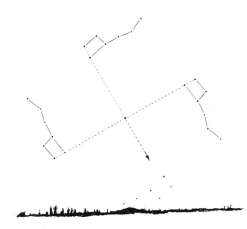

Fig. 7.2. Ursa Major as a celestial clock. This apparent motion shows the rotation of our Earth on its axis.

Fig. 7.3. Organize an astrophotography notebook. Make notes of every photograph you take. Such a log can become your most useful tool.

pointer-stars in Ursa Major, literally the 'Great Bear' and also called the 'Big Dipper' or the 'Plough', which help us locate this 'North Star'.

Today we know that the motion of Ursa Major is brought about by the rotation of the Earth. The 'rump' and 'hindquarters' of the Great Bear have the shape of a dipper with a handle which emphasises the movement of the constellation. Ursa Minor, the 'Little Bear', has a similar shape. Let us re-examine the calendar aspects of the summer and winter Bears more closely.

For most observers in the Northern Hemisphere, the Little Bear and Great Bear are in the sky all the time; the only thing which *seems* to change is their position as it relates to the time of the night and the season of the year. If, for example, the 'handle' points up after sunset, it will point down before sunrise. This apparent motion is a manifestation of revolution of our Earth on its axis.

The same changes in position of the 'handle' can be noted over a period of half a year. If you observe it at the same hour but half a year apart, you can witness the revolution of the Earth around the Sun. Our first photographs will readily prove all of the above and, as we take them, we will establish certain routines which should be maintained right from the start.

Before you set out on your first photo session, organize your astrophotography notebook, about 13 × 20 centimeters in size, and divide the space approximately as shown in Fig. 7.3. This log book can be a smaller ledger which has preprinted vertical lines. It should definitely have a hard cover because it will be used out in the open, sometimes under damp conditions. The principal entries concerning film, date, weather and Moon can usually be made in advance. Notes regarding individual exposures, their duration, time started and ended, are best entered by flashlight with red cellophane over the lens to reduce glare and to keep your eyes dark-adapted.

Use one or two pages per film and leave room because you should not only note data about the pictures as you take them, but you will also want to record additional observations after the film has been developed, concerning the outcome of certain exposures and your own ideas for future work. Very often the photographs taken on a given night are reshot on a following date to improve results after the notes are reviewed with the finished pictures in hand. Best results with specific camera/lens/film combinations can soon be standardized for a particular location. These will allow one to concentrate on a program of familiarization with the sky when most shots are going to be predictably perfect.

The time has come to attach the 'C' clamp camera-holder to a firm base. Use a sturdy tripod if available, but don't buy one; it's much more costly than a 'C' clamp base and not suitable for the telescope you may end up purchasing. Screw the cable-release into the shutter-release socket. (This inexpensive device must have a locking screw for time-exposures.) Point your camera at the North Star, Polaris, or at least in the general direction of this target. Make sure the lens cap is off and double check the 'B', the infinity, and the largest aperture settings.

The important moment is at hand. Press the cable-release and lock it in the 'open' position. Look at your watch with the red-shielded flashlight and note the time in your log. Set a household timer to ring in 5 minutes and relax with your binoculars in a comfortable deckchair or on a blanket in the grass. You have just embarked on an exciting journey to the stars and have set out on what can become your most meaningful and thrilling adventure.

It will be of utmost importance that you start taking notes with your very first shot because it is entirely possible that even your starting photograph will yield unexpected and scientifically valuable results. You might capture a meteor or even a fireball. Then again, an exploding star may be discovered by an observer somewhere in the world tomorrow and your photograph will happen to contain a precious pre-discovery record of it. The 'impossible' and the unexpected do happen and your camera may be the only one in the world pointed in the right direction at the exact moment when astronomical history is being made.

While your camera is accumulating starlight, view the heavens with or without binoculars and start to familiarize yourself with the constellations that surround the Celestial North Pole. Consult one of the many star maps available in star atlases or in the wonderful astronomy publications (see Appendix) which usually show the night and pre-dawn skies of any given month. You will be amazed how quickly you will begin to recognize specific stars and their relationships to one another as they form constellations.

When the timer rings, close your shutter and again enter the time in the log. Next, advance the film and try a 10-minute exposure – and later a 20-minute shot. Record them conscientiously. If there is no noticeable condensation in the air, you may want to leave the camera out and open overnight, setting it before you retire, but turning the f ring to f/16 to let in less light over a longer period of time.

To achieve optimum results with all-night exposures, set your alarm clock for early next morning, *well before dawn*, so that you can close the shutter and save the precious multiple-hour record. If you plan such an all-night 5- or 6-hour picture-taking session, you will find some simple precautions very well worth taking. The first is to put a blank unused frame between your last timed evening exposure and your all-night shot. To do this, simply expose one picture for just a brief second and advance the film again. Make a record of even this pictureless exposure (spare) frame so that your log will show it. Otherwise you may wonder, later on, how that void came to be on your film. The reason for this spacer frame is that you may not hear the alarm clock before dawn the next morning and may oversleep. In that case the morning daylight will enter your wide-open camera. It will not only bleach out the frame in the focal plane, but 'leak' into the two adjoining ones. Such light will definitely spoil at least the edge of your carefully timed final exposure of the night before. Your second preparation for long-exposure work should anticipate early morning dewing, which can often occur even though the midnight air seems dry. The following should suffice: wrap your camera in a translucent polyethylene bag with the lens towards the opening. Cut a tiny hole near the base of the housing for the mounting screw. Put a rubberband around the lens to hold the polybag in place, sliding the band as far forward toward the rim of the lens as possible with just the lens outside the cover. You can place a hairdrier nearby to play warm air over the lens to keep moisture off. Take care not to change the camera settings while doing all this. Now the camera with its 'raincoat' is considerably less vulnerable to a coating of dew, or even a mild shower. Of course, you will certainly want to check the weather report before deciding to leave your equipment out overnight.

It is well worth getting into the habit of opening and closing the camera in a way which will prevent picture blurring caused by shutter movement. Prepare a 30 centimeters × 30 centimeters piece of black cardboard (painted black if necessary). Hold this shield in front of the lens while you open the shutter with a cable-release. Wait 2 or 3 seconds for any camera vibration to die down and then swing the protective cover away from in front of the lens. Reverse the process when you close the shutter. In the old days, street photographers used their hats to achieve the same results with their slow paper-negative concertina-shaped boxes. This method of avoiding blurred pictures was sometimes called 'the hat trick'.

The cardboard allows another trick to be performed. Try one of the longer exposures aiming the camera at a constellation you may already know: Ursa Major (if visible) the 'W' of Cassiopeia, Orion, or the Northern Cross. Hold the cardboard over your lens and open the camera with the cable-release. Now remove the shielding cardboard from in front of your lens for about 20 seconds, then bring it back once again to stop light from entering. Do not close the camera yet but block any more light from entering by holding the cardboard over the open lens for one minute. Do not touch the camera. Next, remove the occulting piece of card and continue with the exposure for a further 8 minutes. Shield the lens once more while you close it with the cable-release. The resulting picture of star trails will show you the constellation of your choice with almost round star images which were formed in the first minute of the split time-exposure. The trails will allow you to establish easily in which direction the Earth was turning because you will be able to tell the beginning from the end of the exposure. The cardboard also comes in handy when you want to shield your camera temporarily while airplanes fly into the area which you are photographing.

To summarize, here is a 12-point checklist which may be of help:

1. Load camera with fastest available film. (Enter film data in log.)
2. Clamp camera securely to a firm base or attach to tripod. (Place in polyethylene bag to protect against dew and use the hairdrier to keep the lens dry.)
3. Aim in the direction of the North Star or constellation of your choice.
4. Set time on 'B' (Bulb) setting.
5. Set focusing ring to infinity.
6. Set aperture ring to widest possible opening for the lens used.
7. Attach cable-release to camera and start exposure by pressing and *locking* the plunger on the camera release. (Or use 'Hat-trick'.)
8. Record all data in log: film, camera settings, start of exposure, Moon if any, weather information, dew conditions, etc.
9. Do not touch the camera during exposure and keep all light away from it.
10. Close camera shutter after desired time. (Use 'Hat-trick' again.)
11. Enter closing time in log and add whatever final observations you may have: clouds passing during exposure, airplane, satellites or shooting star crossing field of view, etc.
12. After removing the film from the camera, if you have it processed by a lab be sure to mark your rolls as follows: 'Important Note. Star Photographs. Normal processing, please.'

Some final thoughts. The reason why I keep photographing the night skies for so many hours is that the lens is a patient eye and film is an untiring and faithful scribe. Together, they can keep unfailing vigil for you anytime. All you have to do is to set up the camera and open the lens to the wonders of the constantly changing Universe.

Many of the events which occur deep in the vastness of space may be beyond the reach of man's most sophisticated photorecording equipment. Even the largest professional instruments are governed by the limitations which our atmosphere imposes on Earth-based telescopes. But there are many happenings which go unrecorded simply because relatively few regular photographic 'patrol' programs are conducted. One automatically assumes that professionals are doing this kind of work constantly. This just is not so.

The major observatories are not, as a rule, engaged in anything but specific research in relatively tiny areas of the heavens. If one compares the celestial scene to one of those huge domes which cover and protect giant telescopes, then the area which one professional astronomer may research, sometimes for weeks on end, will be a region no larger than one tiny rivet on the huge aluminum dome. Of the 41 000 square degrees which comprise the heavens, scientists may concentrate on an area no larger than a tenth of a degree square at a time.

To the amateur, all the heavens are open for study and photograpy. The standard camera 50-millimeter lens will take in a field approximately 24 degrees high by 36 degrees wide, which multiplies out to 864 square degrees. For this reason anyone with a camera could map the entire heavens surrounding our globe by taking a total of no more than 50 exposures. Just as no fish, big or small, can be landed when trolling the vast oceans unless the baited hook is in the water, no celestial events can or will be recorded unless the loaded camera is open.

Meteors are only a small part of the possible astronomical sweepstakes. Shooting stars may be a common sight when observed accidentally on a summer's night. They become precious when caught on film. Giant fireballs are like big marlin, seldom seen but sometimes caught, not just by the skilled fisherman who spends countless summers entering marlin fishing tournaments, but by the lucky visitor to Hawaii who sets a new world record on his very first outing. Yes, there are huge fireballs and they flash through the firmament unseen by most in the hours between midnight and dawn. You don't even have to hold the fishing pole. Just have your hook and line in the water and spend the night 'trolling for fireballs'. There is much more to catch: novae, supernovae, those totally unpredictable stellar explosions which have happened hundreds or thousands of years ago and whose dazzling light is underway to us even now only to blaze forth somewhere, anywhere, and go at first unnoticed. It may be a day or two or even a week before the full-blown nova is discovered by a discerning observer, but it may already be etched on your film on a frame which seemed second-best.

You can readily be the recorder of such a momentous event before the actual discovery is ever made; more yet, without any formal training in astronomy or without knowing the name of a single star, you can become a blinking astronomer (see Chapter 15) – it will not require costly equipment, just curiosity and the discipline regularly to go out there and open your camera to the sky. Perhaps you'll only want to take one picture a night, one film a month. Possibly time and budget will allow you five pictures a night (a film a week). You can plan to do your photography in the evening after dinner or before sunrise or – with equal success – while asleep.

I was shooting for meteors and in the process – quite by accident – recorded an astronomical happening, the like of which had never been photographed before. The pictures showed a nova that had not yet been discovered when the 13 photos were taken of an explosion blazing forth in the night sky, emitting light and energy approximately 400 000 times as intense as that of our own Sun.

So, set up your camera, take a few shots, then open the shutter and retire to bed. You may dream of your own Lucky Star waiting to explode before your camera's patient eye or of meteors crisscrossing the territory which you have staked out for yourself. Perhaps you will discover a comet.

This writer was sound asleep while the unprecedented series of photographs was taken. It recorded Nova Cygni frame by precious frame in the actual process of exploding.

Fig. 8.1. With only a small camera it is possible to take some stunning photographs of the Moon and other celestial objects. This photograph of the crescent New Moon and the planet Mercury was taken at Walden Pond in Massachusetts. The zero-magnitude planet, located a few degrees south of the Moon, was about as far from the Sun as it ever gets. The Moon, only 28 hours past official New Moon, is illuminated by earthshine, the sunlight reflected from what would be a nearly full Earth as seen from the Moon.

CHAPTER 8

Where to look, what to shoot

After your first few rolls of films have been successfully exposed and developed, you will want to experiment, shooting all the fascinating objects in the sky – constellations, planets, stars of special interest, comets, meteors, and so on. That is what this chapter is about. We will assume that you have some means of rigidly mounting your camera, whether it be by propping it up with an assortment of wooden blocks or screwing it onto an expensive chrome-finished tripod with an elaborate pan-head attachment. In a later chapter we will discuss what can be done with an ordinary camera that has been put on a motor-driven mount designed to compensate for the Earth's rotation. In either case, you will be able to aim for the more exciting aspects of astronomy and to take pictures which will allow truly significant scientific observations, or even a discovery. When you think of it, to be able to do these things from your own backyard and with simple equipment that you probably already own, or can borrow, or can build, or can buy for not too much money, is easily overlooked in this age of super-sophisticated science and technology. The excitement and satisfaction of doing meaningful astronomy or making an important scientific discovery will repay, many times over, the small amount of effort and expense needed to get underway. Are you free tonight?

The Sun and the Moon

Starting at the bright end of the scale gives one the advantage of being able to use short exposures. Successful slides and snapshots could be made of the Sun, but we strongly advise you against photographing it except when it is very near the horizon (see page 36). A camera lens is a good 'burning glass' and, aimed at the Sun, it can ruin the inner workings of your favorite camera. You can also ruin the inner workings of your eyeball by viewing the Sun directly through your camera viewfinder. Again we say, do **not** take the chance!

Pictures of the Moon taken with an ordinary camera are small in scale. If you have a camera with a standard lens, remember the words in Chapter 4 and concentrate on lunar snapshots with interesting foregrounds or cloud patterns. With a telephoto lens, you are better off and can record lunar landscapes with an exposure of a fraction of a second. For a start, use the recommended exposure for 'Cloudy Bright (No Shadows)'. You will be impressed with how much detail you can record. For example, if the focal length of your telephoto lens were 364 millimeters then according to Table 4.1 the focal scale would be 10 minutes of arc per millimeter. The moon's angular diameter is about half a degree or 30 minutes of arc, so on your film the Moon's image would be 3 millimeters across; not very large but 'fine-grain' film makes it possible to enlarge greatly the result, at least 30 times or so, which means that your final lunar picture could be 100 millimeters in diameter.

As for the smallest feature on the Moon that you could pick out, there are limitations, set by the granularity of the emulsion, to recording detail. With fine-grain film the smallest crater that could be revealed would have a diameter of about 10 kilometers, the distance corresponding to the size of a single grain in the film, or about one-hundredth of a millimeter.

Another consideration: what is the resolving power of your long lens? If its f-ratio is 7.2, then the lens diameter – the diameter at the smallest point, not on the outside – is 364 millimeters divided by 7.2, or approximately 50 millimeters, close to 2 inches. Dividing 2 inches into 4.5 tells us (see Chapter 4) that the resolving power of the lens is 2.25 seconds of arc. The Moon's diameter is a little over 3400 kilometers, and its angular size is a little under 2000 seconds of arc. If you work through the numbers, the angular scale of the Moon's surface is about 1.9 kilometers per second of arc, so that your lens should be able to resolve a 4-kilometer crater on the Moon. In other words, the film capability cannot match the resolving power of the lens, at least on nights when the air is steady enough to permit this resolution. In most places near sea level, one second of arc is considered to be 'good seeing'; 2.25 seconds would rate only as 'fair'.

Finally, what about the Moon's motion through the sky? Will that appreciably blur the picture? On average, it takes a little over 12 hours for the Moon to go from the east to the west horizon: 180 degrees in 12 hours is 15 degrees per hour, or 15 seconds of arc per second of time. During an exposure of one-fifteenth of a second, the Moon moves just about one second of arc. Fine-grain film is slow and you might have to use a one-eighth of a second exposure, which means again that 2 seconds of arc would be your resolution considering just the Moon's motion.

Fig. 8.2. (overleaf) The most awe-inspiring sight in all the heavens is the totally eclipsed Sun. This photograph was taken in June 1983 when a solar eclipse passed over Java and some other islands in the South Pacific.

With all these factors – film capability, lens resolution, seeing, image motion – working against you, not to say anything about lens quality, steadiness of the mounting, correctness of the focus, or a little dew on the lens, you will find that a 15-kilometer crater is about the smallest discernible feature on the lunar surface. But there are hundreds of craters and mountains larger than that and you will be amazed with the final result. Of course, with a bigger, faster (and more expensive) lens, and with your camera attached to a motorized 'equatorial' mount that follows the Moon's motions, you can do better. Modest beginnings leave lots of room for improvement and plenty of chances to experiment.

For an interesting special bonus, photograph the Moon during one of those occasions when it passes close to or occults a bright star or planet. A telephoto lens of 300 millimeters focal length should just be able to capture the rings of Saturn, the phases of Venus, and possibly even the cloud belts of Jupiter.

There is one outstanding exception to our strong words against photographing the Sun, and that is an eclipsed Sun (Fig. 8.2). A total eclipse of the Sun is, to my mind, the most spectacular celestial event of all. If you have never seen one, make every possible effort to be in the right place at the right time, even if it means traveling great distances. History and legend are full of instances where solar eclipses have contributed to winning battles, have helped overthrow governments, or have saved men from execution. To see a total eclipse of the Sun is to leave you with an unforgettable memory of an experience which no written description, however vivid, can provide; there is far more to it than just seeing the Moon block out the Sun. A series of photographs can record the visual impression of the event and will leave you with an accurate momento of the awesome happening.

One can get beautiful as well as scientifically useful photographs of the delicate coronal streamers and brilliant cloud prominences of the solar atmosphere when the main body of the Sun is blotted out by the Moon. Because the surface brightness of the inner region of the corona is hundreds of times that of the petal-shaped outer area, almost any exposure time will produce superb pictures, and to get the most complete record a wide selection of exposure times should be used – from one-hundredth of a second or less, to a few seconds. Because the Sun, like the Moon, moves less than one per cent of its diameter across the sky in a second, a 2- or 3-second exposure will show little blurring. Kodak publish a pamphlet giving more detailed instructions, including photography of partial phases, and when you prepare your own eclipse expedition, you should obtain this publication (Publ. No. 5–72). Ask for one in your local photostore.

A lunar eclipse (Fig. 8.3), also a fascinating event, can be seen by a sedentary sky-watcher about once a year on the average. Over half the world gets a chance to see the Moon during the hour or so that it takes for it to pass through the Earth's shadow. Here, the object should be to get good photographs of the uneclipsed portion of the Moon, and of course the exposure times are no different from those which one normally would use around the time of Full Moon (the phase of the Moon when eclipsed). As we indicated before, the exposure times correspond closely to those listed in the film package information sheet under 'Cloudy Bright (No Shadows)', but bracket your exposures. One can and should practise on the uneclipsed Moon when it is full one or two months in advance of the eclipse.

Stars

In a pollution-free atmosphere on a crystal-clear, moonless night, far from city lights, the sight of a black sky filled with an overwhelming number of stars is awesome. Observing out in the country or up in the mountains is far more rewarding than from the backyard of a city-dweller's home. Still, much can be achieved in the way of good and useful photographs of stars and star fields, even from city or suburban sites. Dark skies permit long exposures showing to their best advantage star trails (Fig. 8.4) – circular around the North Star (or the South Celestial Pole), parallel near the Celestial Equator where constellations like Orion, Virgo and Pisces are located (see Chapter 13 for more information).

After your experiments in recording star trails and constellations have been successfully completed, look next to variable stars. Not only do they provide fascinating subjects to photograph, but many behave unpredictably and need to be monitored so that professionals can turn their telescopes, equipped with elaborate apparatus, towards these special stars when they start to behave peculiarly. There exist several active, energetic organizations established solely for the purpose of collecting and analysing your observations, and they will provide lists and finding charts of variables. Interested astrophotographers should write for further information and membership application to the American Association of Variable Star Observers, 187 Concord Avenue, Cambridge, MA 02138, USA. First practise by making star-trail photographs of some of the variable stars listed in a star atlas (see Fig. 8.5). Try to catch one of the short-period variables changing brightness rapidly during the course of the exposure, and from your photographs you can construct your own light curve (a graph of magnitude *versus* time; see Fig. 8.6). To follow the complete eclipse of one star by another it may take a number of exposures spaced hours apart.

Fig. 8.3. The partial phases of a lunar eclipse can be photographed with a small stationary camera, or with an ordinary camera mounted on a telescope, which is how this picture was taken.

A word about comparison stars: the AAVSO has published an atlas of the more important variable stars over the entire sky, and on these charts are marked a number of stars of constant brightness together with their magnitudes. From these standards you can easily make brightness estimates of variables with an accuracy of 0.1 or 0.2 magnitudes. Every serious variable star observer or 'blinker' (see Chapter 15) should have a copy of this valuable all-sky atlas.

Most variable stars are of three types: eclipsing, pulsating, and cataclysmic (or eruptive). The first type, the eclipsing variable, is usually highly predictable and consequently the best on which to start. As one star begins to be eclipsed by a fainter one orbiting around it, the brightness can change so rapidly that within a matter of minutes fading can become readily apparent. (Subsequent brightening will occur equally fast.) The whole stellar eclipse can take place within a few hours with the change in brightness amounting to a magnitude or more. Periods between eclipses are typically measured in days, but the full range is truly remarkable – from a few tens of minutes to dozens of years. As you can see from Fig. 8.6, either total or partial eclipses can occur; the type of occultation can be judged from the shape of the light curve.

Members of the AAVSO concentrate primarily on the second type of periodic variable, long-period pulsating stars, or Mira stars, none of which repeat their changes in exactly the same way. Again the range of periods is great, going from several months to many years; most of the stars on the AAVSO list go through their cycle in approximately a year (Fig. 8.7). Their full range in magnitude is usually five or ten magnitudes indicating that these stars go through size oscillations which are truly enormous. (Remember: a range of five magnitudes means a 100-fold change in brightness.)

With persistence, and luck, you can discover an exploding star, a nova (plural: novae) or perhaps a supernova. These cataclysmic variables, which can increase in brightness by many

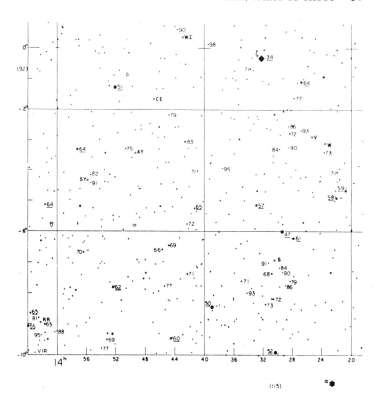

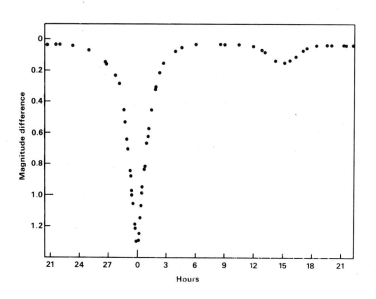

Fig. 8.4. (right, above) The Variable Star Atlas of the American Association of Variable Star Observers shows stars in the entire sky down to about ninth magnitude and gives specific magnitudes for a number of stars in the vicinity of each variable star. A most useful atlas!

Fig. 8.5. (right) The light curve of this eclipsing binary star system (AF Geminorum) exhibits a dramatic primary eclipse, but the secondary eclipse is barely detectable. These observations were made by Carlson Chambliss who chose to compare the brightness of AF Gem with a nearby star. He was a visiting astronomer at the Kitt Peak National Observatory which is operated by AURA, Inc. under contract with the U.S. National Science Foundation.

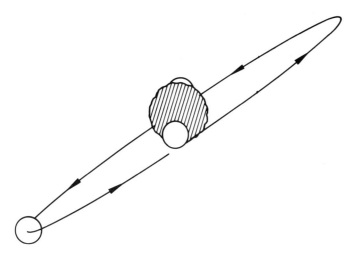

Fig. 8.6. How an eclipsing binary star system would look from close up. The primary eclipse occurs when the small bright star passes behind the larger but fainter companion star. The light curve of this system would look very much like that shown in Fig. 8.6.

magnitudes in one day, are truly explosive stars which nearly always are totally unpredictable. Perhaps a half dozen novae are discovered in our own Milky Way Galaxy system each year, but most of them do not reach naked-eye brightness. In August 1975 Nova Cygni, first spotted by an amateur astronomer in Japan, rose to an impressive second magnitude, the brightest 'new star' in over 30 years. In Chapter 9, you can read about what one amateur astrophotographer was doing on the night that it went off and how he obtained one of the truly unique and scientifically valuable records in the annals of astronomy – and almost lost them.

An average nova increases in brightness 100 000 times (12.5 magnitudes) in a little over a day, finally reaching an absolute magnitude (apparent magnitude at a distance of 32 light years) of -7 or -8. A supernova increases much more spectacularly and at peak brightness is seven or eight magnitudes more luminous even than a nova. If, for example, Pollux, the brighter of the two first-magnitude stars in Gemini and just 32 light years away, were to become a supernova, it would light up the Earth with a brightness equivalent to *ten Full Moons*. A much rarer event than a nova, a supernova occurs in our own Galaxy on average once a century according to best estimates. The last one observed was Kepler's Star of 1604 which reached an astounding *apparent* magnitude of -6, bright enough to be seen easily in the daytime and to cast a clear shadow at night.

Remember: we are overdue for our next supernova, not having seen one in over 300 years. Watch for it!

To discover a nova (or supernova) is perhaps the single most valuable contribution an amateur astronomer can make. Many are in fact discovered by amateurs who have taken the effort and patience to scan the skies regularly looking for stellar newcomers. Obviously, the easy way to patrol our Galaxy of stars is photographically, since pictures taken on different nights can be intercompared and stored permanently. Fixed cameras can record trails down to about the naked-eye limit, and you might conclude that one must remember the locations of the more than 5000 naked-eye stars to pick out a newcomer. In Chapter 15 we will describe some fascinating, home-built devices which remove the necessity of memorizing all those star positions and makes sky-surveying easy for all.

Here is a good tip: novae occur in preferred places in the sky. Over three-quarters of them appear within 15 degrees of the Milky Way and nearly a quarter are found in a single constellation, Sagittarius, in the direction of which the center of our Galaxy is located. Therefore, the lazy (and sensible) way to discover a nova is to patrol only the Milky Way regions, especially towards the galactic center – Sagittarius plus neighboring Scorpius and Aquila.

Supernovae also occur with highest frequency near the

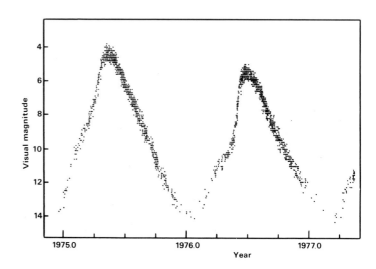

Fig. 8.7. The light curve of a long-period variable. These observations were made by members of the AAVSO; the star is Chi Cygni. Each dot represents a single observation.

Milky Way but remember: a rate of once per 100 years does not make for frequent pay-offs. Professionals or amateurs with large telescopes do better by patrolling other galaxies, especially those sharing membership in large clusters of galaxies. If a nearby star like Pollux were to explode suddenly into a supernova, no finding chart would be needed to pick it up.

The light curve of a typical nova shows that these newcomers remain near maximum brightness for several nights. Even if you miss a night's observing because of clouds or a pressing engagement, you still have almost as good a chance to find it the next night (see Fig. 8.8). However, there are 'fast' novae which fade more rapidly than the nova in Fig. 8.8 and there are 'slow' novae. This rate of drop-off in brightness tells us something about the absolute magnitude of the nova or supernova: the faster the fading, the more luminous the star was at peak intensity. In Chapter 10 we will discuss the reasons behind those stellar eruptions, pulsations and other forms of erratic activity.

Comets

Every so often – perhaps once every 5 years on the average – an impressively bright comet will visit the Earth's neighborhood and provide one of the finest subjects of all for the astrophotographer. Each year a dozen or so fainter cometary wanderers are picked up as they visit briefly the inner parts of the Solar System. Some of them have been known for years and their paths through the sky, their brightnesses, and overall appearances are predictable; others have never been seen in recorded time and provide astronomers with the challenge of calculating precise orbits as well as learning the maximum possible about the visitor in the few weeks of best visibility.

Photographing a bright comet with a stationary camera is simple: aim and shoot using a variety of exposure times. Doubling the exposure produces noticeable results; therefore you might try, for example, 1, 2, 4 and 8 minutes. Increase the developing time by 50 per cent to bring out as much of the faint tail structure as possible (in the case of photolabs, tell them to 'push' your film during developing to double its given ISO rating). Your results should be stunning.

Just when a bright comet will next appear can only be learned by reading current astronomical literature, and even then the comet is sometimes gone before the next issue. Subscribing to the announcement card service of the International Astronomical Union, the *IAU Circulars*, is a most reliable early-warning system. Anyone can subscribe. For more information, write the

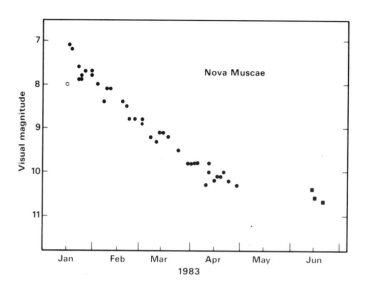

Fig. 8.8. *The light curve of Nova Muscae 1983. Filled circles are magnitude estimates made by this author from his own photographs. The three squares are photoelectric observations also made by WL. The open circle comes from a pre-discovery photograph made in Australia by Eddie Gainsford.*

Editor, IAU Circulars, Center for Astrophysics, 60 Garden St., Cambridge, MA 02138, USA. (See also Chapter 14.) All major new discoveries are reported in these *Circulars*, and all astronomical institutions and many amateurs subscribe to them. The AAVSO also provides comet alerts; for more information write to them directly enclosing a stamped, self-addressed envelope.

The most famous comet of all, Halley's, whose appearances can be traced as far back as 240 BC, makes appearances roughly every 76 years (the exact period varies). Made up of dirty ices and snow, a comet loses material each time it comes in from the deep freeze of outer space and passes close to the Sun. Halley's was a much more awesome sight two millenia ago than it is nowadays.

The greatest thrill of all in amateur astronomy has to be to discover a comet. The fact that quite a few amateurs have several discoveries to their credit proves that finding a new comet is not just a matter of luck by any means. A good search technique, similar to the one used in trying to find a nova, works effectively and can be run concurrently with nova search-programs. For visual searches, one has to remember only those nebulae, galaxies and star clusters which might look like a fuzzy, faint comet;

Fig. 8.9. Comet Ikeya-Seki over downtown Los Angeles, October 1965, photographed by this author with Kodak Tri-X film, f/1.7, 50-millimeter focal length, 32-second exposure. This comet came so close to the Sun that it passed through the corona, nearly grazing the surface of the Sun itself.

'sweeping' the sky with good binoculars or a telescope with a wide field is the usual procedure. Whereas there might be a few hundred or a few thousand stars to learn in a nova search, there are usually only a few dozen or a few hundred nebulous objects to keep track of, depending on how penetrating is your search. However, the 'blinking method' described in Chapter 15 may well prove to be the easiest and most thorough if employed with regularity.

Again, there are favored areas of the sky in which to search for comets – in or near those zodiacal constellations that lie within 30 or 40 degrees of the western horizon after sunset and especially the eastern horizon before sunrise. These are the fields for best hunting but, like novae, comets can occur anywhere in the sky.

Discover a comet and your own name will be bonded to it forever. For as long as the records of mankind persist on this planet, your name will be part of that comet. Start looking tonight. When you have found one, refer to Chapter 14 for information on what to do next. In Chapter 10 we tell a little more about comets, what they are, and where they come from.

Asteroids

Because they almost never become visible to the naked eye and because they appear perfectly star-like except in the largest telescopes, asteroids are perhaps seen to be a less glamorous class of object than comets, novae or supernovae. However, they do have numerous fascinating characteristics: they have been seen closer to the Earth than any other large astronomical object (except the Moon); there are more known than all comets, novae and supernovae put together; and they all bear names given to them by their discoverer or orbit discerner. The huge majority of the thousands of known asteroids, or minor planets as they occasionally are called, stay between the orbits of Mars and Jupiter, but a few do come closer to the Sun and several dozen of them actually cross inside the Earth's orbit.

In 1937, a small camera picked up the trail of a fast-moving eighth-magnitude asteroid, later to be named Hermes. This interplanetary boulder, no more than 1.5 kilometers or so in diameter, passed within a million miles of the Earth at closest approach. It was in fact this near miss which made Hermes appear to move so fast, an illusion analogous to comparing the apparent rate of motion of a jetplane coming in low to buzz an airfield to that of an equally fast-moving aircraft drifting lazily across the sky when seen at a great distance.

Asteroids are picked up most easily by their apparent motion relative to the very distant background stars and for that reason asteroid hunting really requires a camera that can track the slow motion of the stars. A blink comparator (see Chapter 15) is again a useful device for an asteroid search. If you simply want to add photographs of asteroids to your collection of stationary camera photographs, shoot for Vesta, Ceres or Pallas, the brightest of the minor planets with diameters of a few hundred kilometers. Predictions of their positions in the sky are frequently given in the popular astronomy periodicals. Sometimes an expert will predict that a certain asteroid will occult a certain star; a star-trail photo will show the event dramatically.

Unless you discover an asteroid with highly unusual properties, you will not necessarily have the privilege of naming it.

Very often that honor goes to the hard-working calculator of orbits, the person who has to decide which observations to accept or reject. The success of his efforts is proven a year and several months later when the asteroid is again in a good location in the sky to observe. If it is where the calculated orbit says it should be, the asteroid's path has been well determined. Usually after a second observing period the asteroid officially becomes a known minor planet and so receives a name. Asteroid No. 2863, named 'Ben Mayer' to honor my co-author, was discovered by a professional, Dr Edward Bowell at Lowell Observatory; the name was suggested by the expert amateur Phil Dombrowski.

Meteors (and Meteorites)

In order to capture a falling star on film, the main thing to know is not so much where to shoot but when to shoot. In the next chapter there is a table of principal meteor showers and when they occur (Table 9.1). These are the times to expose for meteors, especially when the constellation after which the shower is named is highest in the sky. If you just cannot wait until the next scheduled shower, then concentrate your efforts in the several hours before dawn. Meteors come on all dates at all hours, but at dawn you are on the side of the world facing in the direction towards which the Earth is moving.

Meteors are produced by small pieces of interplanetary debris with sizes measured in millimeters and usually of fragile materials. Colliding with the Earth, they get incinerated as they bore through the upper atmosphere at speeds of many kilometers per second. Meteor showers occur when the Earth encounters a swarm of these fluffy particles left behind in the wake of a comet. (The comet itself need not be visible; in fact, it may be long disintegrated.) In many cases the shower can be identified with the parent comet – for example, the Orionids are provided by Halley's Comet.

Here are a few tips about photographing meteors. If you are primarily seeking these celestial spectaculars, then use a fast film which has a high reciprocity failure, like Tri-X Pan or 1000 ISO color film (see Chapter 6). Remember that the effective exposure time of a meteor picture is a brief fraction of a second with an occasional full-second flash. Shooting at the Celestial Pole produces the best results with a stationary camera, since it is there that the star trails are shortest for a given exposure time. An automatic film advancer, which comes with some cameras or can be easily built if you are clever with gadgets (see Chapter 14), makes it more convenient to shoot during the best times in the pre-dawn sky. If you aim at the constellation for which a shower is named, you will be able to locate the radiant, that point away from which all the shower meteors appear to move (Fig. 10.8). (The radiant effect is an illusion of perspective: all the meteors of one shower are traveling through interplanetary space on parallel paths; like the rails of a railroad, their paths only seem to project back to a point in infinity.)

The faintest meteors you will 'catch' will be of about second magnitude; the brightest will usually rival the major planets and can even be classified as shadow-casters. On occasion you may see or photograph a fireball, a meteor comparable in brightness to the Moon. Something this bright has a small chance of making it all the way through the atmosphere and reaching the ground, especially if it is made of a solid chunk of high-density material. The way to learn if such an object did hit the Earth's surface is either to go looking for it with a search party or to keep alert for reports of falling stones on the radio or in the newspapers. Looking for it yourself is probably the least likely way of finding a fresh meteorite (as a grounded meteor is called) since your photo gives you little idea of its distance from you.

Finding an old meteorite is not much easier because they often look so similar to natural Earth-rocks. If you do beat the odds and by one means or another find a meteorite, report it to the nearest observatory or astronomy department or notify the Smithsonian Astrophysical Observatory at once. While they are perfectly safe to handle, meteorites contain minute traces of radioactivity which die out in a few days or weeks, and immediate laboratory studies of this brief nuclear activity are extremely valuable, as Chapter 10 will reveal.

A final word: you will be paid for any meteorite which you might discover. Maybe not much, but there will be other rewards, especially if the thrill of discovery is what you seek.

Fig. 9.1. The region near the center of our Milky Way lies in the southern summer skies. It makes for wonderful photographs showing myriad stars.

CHAPTER 9
From the wastebasket to the Smithsonian

It can happen to you, just as it happened to me.

I was relatively new to astronomy when I caught my first 'lucky star'. It is 'My first' because I fully expect to catch more. Having tasted the thrill of discovery and reaped some of the rewards that come from recognition, not only from fellow amateurs but also from professional astronomers, one craves for more.

I had graduated from the 'a picture a night, a film a month', had even passed the stage of 'five pictures a night, a film a week' and was concentrating heavily on meteor showers, those wonderful, almost predictable displays of nights filled with shooting stars which happen at certain times of the year (see Table 9.1).

Having had some excellent results with meteor pictures taken during the best (maximum) nights listed in Table 9.1, I soon decided to take as many pictures as possible during the nights preceding and following the maximum dates. I systematically took four pictures an hour, leaving the camera wide open for 15 minutes at a time. At first I aimed at the constellation from which the meteors originated and later arbitrarily at the darkest possible region of the sky. The latter method yielded far better results. (Other meteors which seem to fall randomly throughout the night sky are called 'sporadics'.) The feeling of anticipation which one experiences as one examines with a magnifying glass each new frame on a freshly developed roll of film is a mixture of boundless expectation and suspense. The sight of the needle-like meteor streaking boldly across the star field is a prospect awaited with baited breath, even though one knows that several rolls of film may yield nothing at all.

I had rigged up a Rube Goldberg device (see Chapter 19) which combined a lawnsprinkler–timer and assorted solenoids to advance the film and trip the shutter of an old single-lens-reflex camera. Even though the August Perseid shower had passed, I continued to 'troll for meteors', concentrating on the region of the constellation Cygnus, also known as the Northern Cross, right up to the end of August. In fact I was photographing 15 pictures a night during the short late summer nights when good fortune struck. The drama of the story unfolded gradually. While I attended a 'star party' at the Southern California site in Ojai, where fellow amateurs met on the night of August 31, 1975, almost everybody was talking about a nova which had been discovered the night before by a Japanese non-professional stargazer. 'Nova Cygni' was named after the constellation in which it had appeared. A friend pointed out the bright new star to me where no star should have been. There it was, exactly in the area where I had been photographing all week long, hoping to catch some stray Cygnid meteors. I must be truthful; I would never have noticed the new arrival had someone else not pointed it out to me. By that night it had grown into a star of second magnitude, which put it among the brightest in the heavens.

What I could tell no one on that Saturday evening, when the nova was the subject of all conversation and the object in every viewfinder and telescope, was that I had just that morning developed the week's Cygnus Constellation film which I had taken, searched the negatives for meteors, and, finding none, had dumped three rolls worth of film into the wastebasket. Here I

Table 9.1 *Principal monthly meteor showers*

Name of shower	Dates	Maximum	Estimated meteors per hour when at maximum
Quadrantids	Jan. 1–5	Jan. 3–4	20–80
Alpha Aurigids	Jan. 15–Feb 20	Feb. 7–8	12
Zeta Bootids	Mar. 9–12	Mar. 10	10
Lyrids (April)	Apr. 19–24	Apr. 22	12
Eta Aquarids	May 1–12	May 5	20
Lyrids (June)	Jun. 10–21	Jun. 15	15
Delta Aquarids	Jul. 15–Aug. 15	Jul. 28	35
Perseids*	Aug. 1–18	Aug. 12*	65
Beta Cassiopeids	Sep. 7–15	Sep. 11	10
Orionids	Oct. 17–26	Oct. 20	35
Leonids*	Nov. 14–20	Nov. 17*	10–100
Geminids*	Dec. 4–16	Dec. 13–14*	50

*Major showers. Try not to miss these dates

was with my secret in Ojai, far from a telephone, haunted by the thought of the precious emulsions even now sharing the garbage can with melon rinds and coffee grounds. All I could do was to concentrate on the brilliant celestial newcomer and to photograph it carefully through my largest telescope.

Next morning I went in search of a telephone to contact my West Los Angeles home. 'Have you by any chance emptied out the wastebasket in the den?', I asked my 18-year-old son who was sometimes given to moments of frenzied tidiness and on other days displayed the laziness of a sloth. 'No Dad', he said, 'But I will . . .' 'Don't!', I interrupted, 'Don't throw anything out, just lay the negatives from the wastebasket on my desk emulsion-side up.'

The rest is astronomical history. Since no nova had ever been photographed during its evolution, an as yet unexplained oddity was discovered: instead of getting brighter in each succeeding exposure, some of the negatives showed a temporary darkening before the explosion resumed, making Nova Cygni remarkable not only for its eventual brilliance but also for the rapid and pulsing way by which it grew to magnitude 1.8. The subsequent fading to magnitude 8 was almost as rapid which, as Dr Liller explained, made it a record-breaking 'fast nova' with an energy output of hundreds of thousands of Suns.

Soon my photographs, with proper credit, appeared in many publications and as the noted amateur (and medical doctor) J.U. Gunter wrote, 'Ben was catapulted from the trash can to instant fame' (*Tonight's asteroids*, Bulletin No.29). The astronomer Dr Edward K. Upton wrote, 'The catching of a fast nova, step by step in the act of becoming visible, is without precedent in the annals of astronomy' ('The night of the nova', *Griffith Observer*, November 1975). To cap it all, the photograph which I took at the 'star party' in Ojai was included in the 1977 Encyclopaedia Britannica *Yearbook of science and the future*. Other photos have since appeared in many major astronomy textbooks.

It all happened – and it can happen to you too. To prove it, two other very important 'pre-discovery' photographs of Nova Cygni must be mentioned here; they were taken by Peter Garnavich, then an amateur not yet out of his teens, in Bowie, Maryland. Since it gets dark earlier on the eastern seaboard, Peter, who was conducting a regular 'Milky Way Patrol', was ahead of me by about 3 hours. More proof that 'keeping the shutter open' can bring wonderful rewards to anyone, anywhere. Peter's negatives were also sent to Dr Liller and the first Garnavich picture showed the Nova at magnitude 8.4. The second shot, taken only 3 minutes later, registered at magnitude 8.2. My own photos began with magnitude 6.2 and through the 13 pre-discovery shots increased in brightness to magnitude 3.1. That was the point at which the Nova was first actually seen by Kentaro Osada in Japan.

Anyone can do what Peter and I did. The more people who aim their cameras at the skies, the better are the chances for amateur contributions to the professional field of astronomy. Even the smallest findings can become of important benefit when studied with other similar material. The pride of having made a contribution is its own reward. Nova Cygni, 'My Nova' taught me several important lessons. Here they are:

1. **Never throw away any exposed film.** Every frame, however insignificant it may seem at the time, may contain a treasure which will not be unearthed until a month or even years later when some research or discovery made by someone else will cause you to look at your own pictures again, perhaps to find what may be no more than a tiny but important ingredient of information which will be of use to scientists. If your first interest is meteors because of their exciting and brilliant though unpredictable appearance, you may not even print every negative. You will be satisfied merely to examine the film strip with a magnifying glass over some kind of a light table which you can easily build with a piece of 'milk glass'. Not only had I decided not to print the negatives of the last week of August but I had actually dumped them. In retrospect I cringe at all the other films I have thrown away. Camera shops offer books of glassine envelopes or little files even, which make the orderly storage of negative material easy. Although it is not to be recommended, the film can even be rolled up and stored in the original container after it has been properly numbered, dated, labeled and identified. Whatever you do, keep all films and keep all records. Number the films and key them to the numbered records. Nowadays I expose the first two frames or the last by shooting a blackboard on which I chalk all the pertinent information about the camera, lens, f/stops, etc., so they automatically become part of the film.

I have said it before; without records the pictures lose much of their value. Dates, exact times and lengths of exposures are the three most important pieces of information needed to fit your work into what could, any day, become a Smithsonian jigsaw puzzle of data from different sources, literally from all over the world. Don't let this give the false impression that everybody is constantly photographing the heavens – quite the contrary. It is just that when a celestial event occurs those lucky enough to have had their cameras trained on the particular area will eventually find out and will then be able to submit their own material.

2. Join an amateur astronomy club. Without a continuing exchange with other amateurs, I would never even have known about Nova Cygni and my films would surely have gone out with the garbage on the following Tuesday night. Don't be discouraged at first to find that many of the club members have telescopes; you may be amazed to find how few of them are into photography and think it is a complex science. Your expertise from the start may be greater in this relatively unexplored field than you can imagine. From the few club members who work in astrophotography, you will learn much through conversation and general sharing of information. Keep your eyes and ears open when someone shows a picture which you like. Note down the type of film used, how long the exposure or how fast the lens.

3. Subscribe to at least one astronomy publication. There are several good monthlies devoted to astronomy. A list can be found in the Appendix. They report important celestial events in advance, announce expected meteor showers, passing asteroids and other predictable occurences. They also list accurate dates and where to look. Skycharts are provided which show the stars for the current month of the year. These maps will assist in the location of the principal constellations, point out the planets that are visible and generally help you find your way in the sky. An increasing amount of space is being devoted to astrophotography so that here too you can learn from the experience of others. The advertising section's classified columns will enable you to make decisions on the type of equipment you may, in time, consider for purchase and just looking at the 'For sale' column will make you wonder why one type of telescope is always offered where others are being sought much more often for purchase.

4. Keep that camera pointed at the sky and keep it open. In this fundamental recommendation lies the entire purpose of the odd chapters in this book. When we say that astrophotography is a relatively new endeavor, I should explain a little further: many would-be astrophotographers are put off this hobby by the belief that there is nothing new to be done in the field anymore. Newspaper stories reinforce such assumptions when they speak only of discoveries in distant space being made by noted astronomers using fantastic million-dollar instruments. Photographs taken by government-funded satellites seem to have made Earth-based astronomy into a useless pursuit. This simply is not true.

It can be re-emphasized that the first pre-discovery photo of Nova Cygni was taken by a 17-year-old amateur in Maryland, while the first recorded visual discovery was made by an amateur in Japan 12 hours later. Even though the nova was 'independently discovered' by many professionals subsequently, its discovery is credited to Kentaro Osada because he first reported the nova's appearance.

A very large percentage of comet discoveries has always been and continues to be made by devoted amateurs. In Japan, comet-searching (and -finding) seems to have been raised to an art form. In many instances only binoculars are used which, combined with a good knowledge of the heavens, have produced spectacular results. Unlike starlovers, who started their astronomical pursuits while very young when their memories were keen, this writer does *not* have a good knowledge of the heavens. Even today I have problems finding more than the most conspicuous and 'easy to find' constellations in the sky. That is why in Chapter 15 we shall discuss what may well be the most important forward step that can be taken by the amateur astrophotographer in contributing to the science of astronomy. Much of what we discuss will be aimed towards 'blinking', a most simple method to compare optically a photo taken, say, last April with one taken in May or for that matter anytime before or after.

The Japanese comet-searcher who patiently views the evening sky after sunset and is out before dawn with his binoculars may know the sky and exactly 'what' should be 'where'. Our method is much less demanding, requiring no more than an ongoing program of photography, combined with an almost mechanical comparison of old and new photographs in indoor comfort on rainy weekends or cloudy evenings. Employing modern phototechnology which has put cameras and slide projectors into millions of hands and combining it with some old fashioned hobby-like tinkering, makes fantastic breakthroughs possible for the amateur. These will result in important spin-offs for the astrophysical sciences. Anyone can play and stake out whatever territory they choose to photograph in the boundless reaches of the sky.

You may wish to patrol large areas with wider-angle lenses which capture bigger pieces of sky. Then again you may want to become a specialist in one, two or three smaller areas of stars. By the time you will have purchased your telescope you will know these areas so well that, depending on their location, it will be easy for you to record the path of a planet as it wanders slowly through your territory. There are hundreds of asteroids to watch. These tiny celestial wanderers travel in their own orbits around the Sun. The brightest ones are Ceres, Pallas, Vesta, Juno and Eros. Those

66 THE CAMBRIDGE ASTRONOMY GUIDE

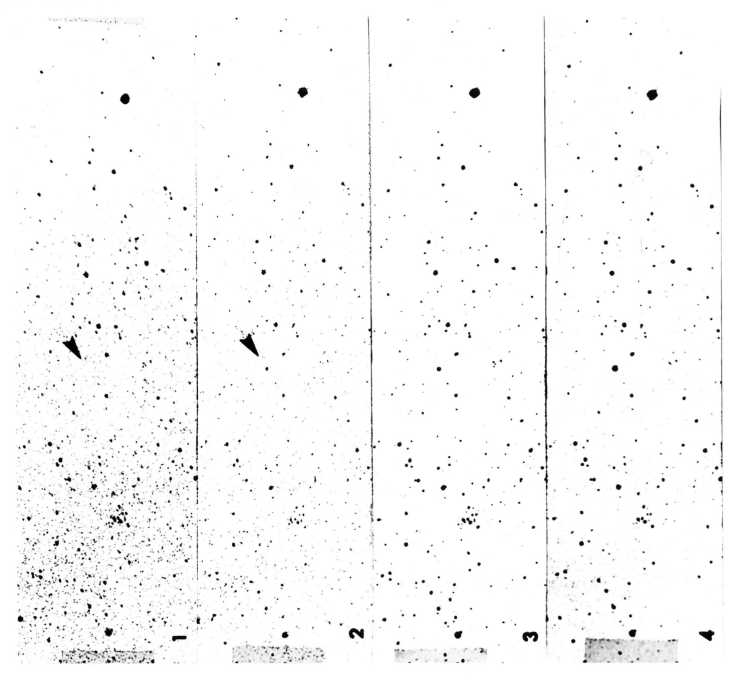

miniature planets range in size from a few hundred kilometers to less than one kilometer in diameter and may make a hurried or leisurely trip through your domain. You will be able to photograph their movements on successive nights (sometimes in successive hours) and observe the reflected sunlight of such visitors as they hurl through your empty space.

Anywhere you aim, anytime you shoot, the very real possibility exists that one pinspot of light which was not there the night before may be your own lucky star just beginning to explode or an uncataloged asteroid. If it is a comet, never before seen, that heavenly body may bear your name for all time and even though it may not come this way again for another thousand years, it will still be known as yours and yours alone. No longer need we just watch and stand on the sidelines in awe but we can now participate and record for posterity all that we can actually see, and more. It does not matter if the initial recording of a comet is taken by your camera and the 'seeing' is not done by you until the following weekend. If you are the first discoverer, the discovery will still be credited to you if you report it properly.

One last word: from most astronomy publications you will learn of comets, old and new, of interesting combinations where a particular planet may be very near the Moon or may 'occult' (block the light of) a bright, well-known star. Unlike the big space-spectaculars, which are written up in the media, the routinely photographed, other, slightly less imposing celestial events are still newsworthy and of great interest to many. Television stations and papers ranging from the *Los Angeles Times* to the *Santa Monica Evening Outlook* have used my amateur photographs. Some of the media even offer to pay for the use of such material but if you're like me the satisfaction of seeing my work presented to a large audience will fill you with a pride and satisfaction which cannot be measured in terms of money. In fact you will find yourself reading up on your subject gaining the expertise which an editor almost expects you to have. To that editor you are the expert. Give as much information with your picture as you can but be sure it is all correct; it is better to suggest that they contact the nearest observatory for more definitive data to go with your photograph than to make an error (see Chapter 14).

It has been said that an expert is someone who knows 10 per cent more than the next person. Similarly it has been stated that some people are luckier than others. Expertise and luck, however, cannot be reduced to such simple phrases. Scientific knowledge can only be increased by setting in order the facts of one's experiences. The beginning amateur will at first lack adequate experience and must go out to seek and accumulate it. Luck is not something that 'just happens'. In astrophotography, the streak of luck which records the streak of a fireball or the trail of the comet often involves the open shutter of a camera. Behind that patient eye stand disciplined amateurs. We wish them well.

Fig. 9.2. Part of the series of sequential photographs taken serendipitously from a Los Angeles rooftop showing Nova Cygni 1975 in the process of eruption. Cygnus Region near Deneb. All photographs 35-millimeter lens, f/3.5. Tri-X Film, ISO 400.

Fig. 10.1. Globular cluster, M92, in the constellation Hercules.

CHAPTER 10
Understanding your camera subjects

In the last several chapters we have specified where to shoot and what to photograph, emphasizing that amateurs' contributions can be of truly profound importance. Before we proceed further, we shall spend a few pages describing in more detail the astronomical bodies themselves. Knowing more about a type of object, familiarizing yourself with its behavior patterns and the reasons for the behavior, will add to the appeal of your adventures and improve your chances for a real discovery. Watching a comet develop a tail during the course of a few weeks is a fascination in itself, and seeing a star suddenly brighten can provide tremendous excitement, but sooner or later one wants to know why these things happen. Such is the legacy of intelligence.

Novae, supernovae and other variables

What causes a star to rise suddenly from obscurity to be one of the brightest stars in the sky? What is the process which produces greater and greater amounts of light in such a short time and how is it triggered?

Nova Cygni 1975 is now one of the best observed novae in history, due in part to the well-documented record (rescued in the nick of time from the wastebasket) of its rapid increase in brightness. We will consider it as an example of its class. Observations showed that the star's *color* did not change significantly as it brightened during those few brief hours in August 1975, and this tells us that the temperature remained surprisingly constant while the dramatic change in magnitude took place. The pre-outburst star was 'merely' white hot which, according to the calculations made by astronomers, indicates a temperature of the

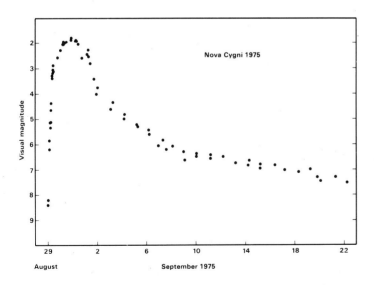

Fig. 10.2. The light curve of Nova Cygni 1975. The 13 points on the night of August 29 lying between visual magnitudes 2.8 and 6.3 were measured by me using Ben Mayer's fantastic series of photographs. The fainter points on that night came from pictures taken by Peter Garnavitch. The remaining observations came from a variety of sources, mainly AAVSO.

Fig. 10.3. A close-up view of what a nova probably looks like. The nova itself results from a nuclear reaction occurring at the surface of a hot dwarf star whose gravity has been sucking gas away from a nearby companion star not unlike the Sun.

Fig. 10.4. This pair of photographs of Messier 100 shows dramatically how conspicuous a supernova (p. 71) in another spiral galaxy can be. On occasions a supernova can outshine its parent galaxy.

hot, glowing gases in the outer layers of around 10 000°C. (For comparison, our own star, the Sun, which is yellowish-white hot, has a surface temperature of 6000°C; the hottest stars known have temperatures at their surfaces of nearly a million degrees.)

The only known way in which a star can become brighter without changing temperature is by growing larger; spectrographic analysis of Nova Cygni supports this conclusion. Since a nova increases in intensity nearly a million times, there must be an increase in surface area of a million times, equivalent to a change in stellar diameter of 1000 times. Before it started its outburst, the pre-nova star was small as stars go, no more than 150 000 kilometers in diameter – not much larger than the planet Jupiter. But over the course of a few hours – a few days at most – the gaseous outer layers of Nova Cygni expanded at an absolutely fantastic rate and by the time it reached maximum brightness in August 1975, the star's total diameter equaled that of the Earth's orbit around the Sun. The outward velocity required to achieve this incredible feat works out to be more than 1000 kilometers *per second*, or about 3000 times the speed of sound.

Even after maximum brightness is reached, such a bubble of gas continues to expand and at nearly the same fantastic rate. However, the same expansion causes the material to become less and less dense. Instead of looking like a growing solid sphere, the nova, if it could be viewed from close up, would take on the appearance of a huge, faintly luminous cloud surrounding a dimly seen star. As time progresses, the nebulous matter continues to grow and becomes more rarefied, while the inner star, now fully recovered, regains its original size. Sometimes, within a year or so, large telescopes can photograph the expanding gas shell and astronomers can keep track of its size by direct measurement.

A supernova behaves in much the same way, but everything occurs on a vastly grander scale: the original star is many times heavier than the Sun; the force of the explosion is catastrophic; the speed with which the outer layers expand is many times higher; and a much larger quantity of gas is hurled into the space surrounding the star. In a nova, only a fraction of a per cent of the star's material is blown away; the event can occur over and over again – and probably does. In fact, about a dozen nova-like stars have been observed to repeat their outbursts at intervals of many years. For a supernova, one explosion is all the star can survive; the remains consist of a rapidly expanding, massive cloud of superhot gas and a small dense star that was once the innermost core of the pre-supernova. In 1054 AD such an explosion took place in the constellation of Taurus, and we now see there the Crab Nebula with its faint central star, so dense that all atoms are squeezed into a degenerate neutron mass.

Why does a star suddenly undergo such an explosion? The events leading up to a supernova outburst have to be deduced from detailed and immensely complex calculations of how a star evolves. With computers, one can calculate the effects of the year-by-year ageing process and keep track of how fast the thermonuclear furnace at the star's center converts matter into energy – and what the escaping energy looks like to a distant human observer. It is now fairly certain that only the most massive stars undergo the supernova catastrophe, the result probably of a star overextending itself, becoming so large, as its core converts atomic nuclei into light and heat, that when the supply of atomic fuel runs short the star suddenly collapses.

As for the milder nova explosion, we find, without exception, that the pre-nova star belongs to a close double system, two stars, nearly in contact, revolving around each other in a matter of a few hours. What appears to happen is that immense tidal forces gradually strip the outer layers of gas from the cooler of the two stars. This gas then is free to move around in between the pair of whirling stars. Some mechanism, probably an intense magnetic field, prevents this liberated gas from falling onto the surface of the hotter, more dense companion. It accumulates over the years on a sort of magnetic roof – until it suddenly collapses. Within minutes the stuff of the cool star pours down onto the hot component, producing an intense flare-up of light.

One example of a recurrent nova normally appears as a twelfth-magnitude star, also in the constellation of Cygnus, but once every couple of months it suddenly rises to eighth magnitude only to drop back after its few days of glory to twelfth magnitude. This star system, named SS Cygni, has been a favorite subject of AAVSO members ever since that association was founded in 1911, and not a single outburst has been missed.

One of the special kinds of variables well studied by the AAVSO, the long-period pulsating star, has in its evolution reached an unstable condition due perhaps to a brief overproduction of energy. Like a giant pulsating balloon, it rhythmically expands and contracts with a complete cycle amounting usually to a few hundred days. Omicron Ceti, known to the ancients as Mira, is the prototype, fading from magnitude 2 or 3 to as faint as ninth magnitude and then returning again to maximum brightness in a cycle of about 11 months. Truly of giant size, Mira-type stars have diameters as large as the orbits of the inner planets of our Solar System. They are 'merely' red hot, telling us that their outer layers glow at a temperature of an astronomically cool 2000–3000°C (see the light curve in Fig. 8.8).

Several other kinds of stars pulsate also, the best known being the *cepheids* (Fig. 10.5), named after one of the brighter examples, Delta Cephei. They differ from the Mira-types by having shorter periods and being two or three times hotter and more luminous; in fact some rank among the most luminous stars in

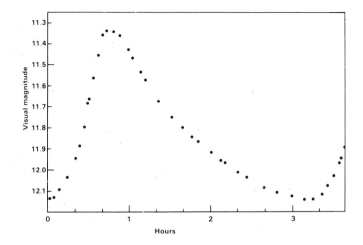

Fig. 10.5. The light curve of a short-period cepheid variable star. This star, XX Cygni, has a period of only 3 hours 14 minutes. Adopted from photoelectric observations made by Michael Joner at Brigham Young University.

space. One remarkable characteristic of the cepheids is that there exists a definite relation between their period and their absolute magnitude. The longer the period, the greater the luminosity.

In the same way that we could calculate how far it is to a distant streetlight, given the wattage of the bulb and its apparent brightness, we can deduce the distance to a cepheid variable. Cepheids serve as luminous standard beacons throughout space with the period of flashes telling us the brightness of the bulb, as it were. Consequently, these variables have been an immensely useful kind of star in establishing distances within our Milky Way, and the distances to other neighboring galaxies.

Another kind of useful star whose light appears variable is the *eclipsing binary* system, where one star regularly passes in front of (and behind) another. As we mentioned in Chapter 8, all types of stars are found in these whirling pairs, and virtually all periods are represented, from approximately an hour on up. In fact, it is the rule, rather than the exception, to find stars in pairs or in groups of three, four or more. The Sun may or may not be counted among this majority depending on whether you would consider its family of planets as legitimate astronomical children.

If the brighter star is totally eclipsed by a larger, much fainter companion, the depth of the eclipse can amount to several magnitudes. If the motion of the stars about one another is rapid, then the change in brightness occurs quickly and dramatically.

Keep an eye on Beta Persei, the second-magnitude star high up in the Northern Milky Way. Every 68 hours this star, called Algol by ancient astronomers, fades by over a magnitude in a little over 2 hours and then recovers, having been nearly totally eclipsed by a larger, but nonetheless fainter, companion.

A word about star names: long ago the brighter stars were assigned Greek letters and Arabic numbers for identification. The brightest star in a constellation is usually Alpha, Beta next, then Gamma, and so on. (Lower case letters are always used: α, β, γ, etc. See Appendix for letters in the Greek alphabet). The naked-eye stars have also been numbered, counting from west to east within each constellation. Therefore, the first star one can see in Orion, moving in from the west, is numbered 1; the last and easternmost is No. 78. A grammatical detail carried over from ancient times is the use of the Latin genitive case in denoting the constellation name. Therefore, Vega, the brightest star in Lyra, is α Lyrae and also 3 Lyrae; Rigel in Orion is β Orionis and 19 Ori, where the constellation name is expressed in abbreviated form.

Variable stars, apart from the bright ones which rate Greek letters (like Beta Persei), are given Roman letters beginning with R for the first discovered, S for the next and so on. After Z comes RR, RS, RT, . . . SS, ST, . . . etc; after ZZ we go to AA, AB, AC, . . ., QY and QZ; and if still more variables are known, we then start counting beginning with V 335. We have often talked about Nova Cygni; that 1975 nova is now recorded as V 1500 Cygni.

To keep this bewildering array of Greek and Roman letters, Arabic numbers and Latin names and cases straightened out, buy a good atlas such as Norton's, or Tirion's.

Comets

In Chapter 8 we wrote of how, every so often, an unannounced visitor journeys in from deep outer space into the inner regions of the Solar System where the Sun and the Earth are located, and provides Earth-persons with an awesome sight, that of a hazy, often gold-colored star from which a tail gradually lengthens as it nears the Earth-Sun system. Telescopes reveal a stellar-appearing *nucleus* in the comet head which must in reality be no more than a kilometer or so in diameter. Surrounding the nucleus is a much larger fuzziness which at the center merges into the nucleus and then gradually fades away to nothing at all many thousands of kilometers from the core. This outer haziness, or *coma* as it is called, often greater in extent than the diameters of the largest planets, develops from a small fuzz-ball of material when the comet is out beyond the orbit of Mars, 2 or 3 astronomical units from the Sun, to its maximum size usually shortly after closest approach to the Sun.

Detailed studies of its behavior have led us to the conclusion that a comet is nothing more than a gigantic dirty snowball, full of ice, dust and volatile gases which gradually melt away as solar energy works its effect. Fred Whipple of the Harvard and Smithsonian Observatories, who first worked out this theory, has called a comet 'a flying cocktail' since many of the volatile materials are nothing more than alcohol ices and sugars that vaporize when close to the Sun.

In the near-vacuum of space between the planets, small ice crystals and dust particles feel the pressure of sunlight as a major force. As this solid material is freed by the melting action of the Sun, the solar radiation pressure gently nudges this comet material away from the outer coma and blows it in a direction away from the Sun, producing the visible tail. An additional source of outward pressure comes from the steady outflow of rarefied gas from the Sun – the solar wind – which moves at high speeds and is able to carry away some of the liberated gases in the coma.

Comet tails can be enormous, stretching in some spectacular cases across the entire orbits of the inner planets, Mercury and Venus, and seen here on Earth extending from horizon to zenith. But as spectacular as they can be, the density of matter in comet tails is extremely low. Although on occasion the Earth has passed right through a conspicuous tail (such as that of Halley's Comet in 1910), nothing has ever been seen to happen. Spectrographic analysis has revealed the presence of carbon monoxide in these tails, but the amount of poisonous gas deposited in the Earth's atmosphere by a large comet is comparable to what comes out of the exhaust pipes of a few dozen automobiles driving along the Los Angeles freeways for a day.

As a comet approaches the Sun, its overall brightness increases rapidly; every time it halves its distance to the Sun, its intensity increases by three or four magnitudes, sometimes more. Comets that come closer to the Sun than Mercury, in some cases passing through the Sun's outer corona, can turn out to be spectacular sights from here on Earth because of their rate of brightening.

Basically there are two kinds of comets: periodic and non-periodic. Halley's (Fig. 10.6) belongs to the first group, returning as it does every 76 years or so to our neighborhood. More typical is Jupiter's 'family', a group of nearly 100 comets which were 'captured' by the most massive planet in the Solar System and whose orbital periods average a little over 6 years. They move in elliptical orbits with a maximum solar distance approximately equal to that of Jupiter and minimum distances varying from less than one to several astronomical units. Because of their frequent close passages to the Sun, these short-period comets tend to be

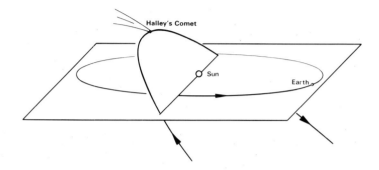

Fig. 10.6. The orbit of Halley's Comet brings this famous visitor up from below the plane of the Earth's orbit – if we assume north is in the up direction. Here we see the relative positions of the comet, Sun and Earth shortly before the comet is closest to the Sun. Sometimes (not in 1986) the Earth can pass very close to the comet as the comet crosses the plane of Earth's orbit on its journey away from the Sun.

fainter than average, having had much of their material melted away over the years.

The non-periodic comets usually put on more exciting displays. Most are probably really periodic, but the time needed to orbit the Sun is thousands of years and when they do come in close to the Sun they are, of course, being seen by civilized man for the very first time. Because they have visited warmer places of the Solar System less frequently than the members of Jupiter's 'family', they tend to be larger and brighter than the short-period comets.

The orbit of a long-period comet has a string-bean shape with the Sun situated just inside one end. At maximum distance from the Sun, thousands of astronomical units beyond Pluto's orbit, the comet moves slowly through the vacuum of nearby interstellar space. There it drifts in its slender orbit for over 90 per cent of the time, deep-deep-frozen, dark and undisturbed. Gradually, after making the sharp distant turn, the comet begins to pick up speed and literally falls in (because of the Sun's gravity) to the Earth–Sun neighborhood where, in a few weeks, it flashes through the inner regions of the Solar System and then heads once again for a slow journey, lasting millenia, through the interstellar deep freeze.

Like asteroids, comets, too, can come close to the Earth on occasion. In April 1983 a small comet, appearing like a tenth-magnitude fuzzy patch, was discovered in quick succession by a satellite telescope (IRAS) and two amateurs, an Englishman,

George Alcock, and a Japanese, Genichi Araki, both of whom did their discovery work photographically. By mid-May, Comet IRAS–Araki–Alcock (Fig. 10.7) approached the Earth to within less than 5 million kilometers and was visible to the unaided eye as a faint, very diffuse patch in the northern sky. Within a few days it faded to telescopic magnitude and could only be tracked with large cameras.

Because Halley's Comet has been so well observed in past appearances, orbit experts can tell us exactly where and when to look for this famous celestial body. Table 10.1 gives positions of the comet for a number of dates during its visit in 1985 and 1986. Also included are estimates of magnitude and tail length, but these quantities depend on unknown factors like the amount of solar activity at the time. Listed finally is the comet–Sun angle.

The origin of comets is still not clearly understood, and because 'uncaptured' comets spend most of their time at tremendous distances from the Earth, it may be some while before we understand comets thoroughly. The prevailing opinion is that at the distance of the outer ends of the long-period orbits there exists a swarm of cometary snowballs left over from the early beginnings of the Solar System, a cloud of material that was unable to condense into a planet far beyond the orbit of distant Pluto. From time to time a star passes near its icy fringe, gravitationally stirs up its members, and starts some of the snowballs rolling, as it were, towards the Sun.

Table 10.1 *Ephemeris of Halley's Comet, 1985–86*

Date (U.T.)		R.A. (1950)	Dec.	Mag.	Tail length	Sun
1985 Jan.	1.0	$5^h 45.5^m$	+12°02′	17.9	0.0°	162°
Apr.	1.0	4 51.3	+14 52	17.4	0.0	63
Jul.	1.0	5 32.1	+18 14	16.1	0.0	16
Oct.	1.0	6 11.5	+20 00	11.4	0.0	95
Nov.	1.0	5 22.5	+21 47	8.9	0.0	137
	15.0	3 58.4	+21 55	7.4	0.0	170
Dec.	1.0	1 06.8	+13 43	6.3	0.4	132
	15.0	23 17.0	+ 3 43	5.9	6.2	89
1986 Jan.	1.0	22 15.9	− 2 29	5.4	9.3	55
Feb.	1.0	21 17.5	− 8 31	3.2	4.0	10
Mar.	1.0	20 27.4	−16 12	4.7	12.2	35
	15.0	19 59.0	−22 28	4.9	15.1	56
Apr.	1.0	18 23.6	−38 35	4.1	13.3	95
	5.0	17 22.8	−44 05	4.0	11.0	110
	10.0	15 25.1	−47 29	4.0	6.5	131
	15.0	13 22.3	−42 17	4.3	2.6	147
	20.0	12 04.7	−32 58	4.8	0.7	148
May	1.0	10 54.8	−18 22	6.1	0.0	129
Jun.	1.0	10 23.9	− 6 34	8.1	0.0	91
Sep.	1.0	11 09.2	− 7 48	12.6	0.0	18
Dec.	1.0	11 40.0	−14 55	14.1	0.0	67

Adopted from the computations of Donald K. Yeomans and Malcolm B. Niedner, Jr. The magnitude and tail length values are estimates based on past performances of the comet.

Fig. 10.7. Comet IRAS–Araki–Alcock passing by the Praesepe star cluster. At the time this photograph was taken, the comet was a mere 5.1 million kilometers from the Earth.

Meteors

On any clear, unpolluted, moonless night, a sharp-eyed observer can usually see from six to 60 meteors (or shooting stars) an hour, depending on the time of night and time of year. Two observers, separated by a few kilometers will see the same meteors but in slightly different directions, proving that they occur relatively nearby, some 60 to 100 kilometers high in the Earth's atmosphere. With precision photographic or radar observations and straightforward triangulation techniques, the meteor paths can be mathematically located in the upper atmosphere. With a rapidly rotating fan blade in front of the camera lens, every photographic meteor streak can be converted into a dashed line, the distance between interruptions giving us the meteor's velocity and even the deceleration caused by the atmospheric friction it encounters as it plunges earthward. Finally, knowing the sizes of the forces involved (mainly the Earth's and Sun's gravitational pulls), astronomers can deduce the original meteor orbit and study the relationship of meteors to other objects in the Solar System.

The results are unmistakable: over 99 per cent of all meteors come from comets. Many meteor orbits can be matched with known comets, and almost all the others travel in orbits typical of those followed by comets in Jupiter's 'family'. The best known and best observed of the meteor showers, the Perseids, are related to Comet Swift–Tuttle, a small comet discovered in 1862. The Orionids come to us along orbits virtually identical to that of Halley's Comet.

Meteor showers can be spectacular if the comet is nearby. Two examples are the Leonids (Fig. 10.8), which become most numerous every 33 years, the period of their parent comet, and the Draconids, which gave superb showers in 1933 and 1946 when the parent comet, Giacobini–Zinner, was close to the Earth.

As we mentioned earlier, passing through the tail of a comet produces no visible phenomena for observers on Earth, but going through the 'wake' does. A comet's tail is composed only of gas and the finest of particles, while the heavier pieces are strewn along the orbit behind the comet. By 'heavy' we mean several grams; at the high speeds with which meteors pass through the thin upper air – typically 25, 50 or 70 kilometers per *second* – the matter of a meteor is quickly and often brilliantly vaporized.

Moving along in nearly parallel paths separated by tens or hundreds of kilometers, meteors of a single shower can be seen all over the sky by one observer, but these parallel trajectories produce the illusion, as do railroad tracks, of a single radiant point. If you plot the paths of the members of a rich shower on a map of the constellations, you can quickly and accurately locate the radiant. Better than visual plotting, use photography. Aim your camera at the radiant and follow the suggestions given in Chapter 8. With a little luck, a rich shower, fast film and a good lens, you may capture a half dozen or more trails all of which, when extended backwards, will very nearly intersect in one point, the radiant, in the constellation after which the shower is named.

The random or sporadic meteors which one sees on non-shower nights are believed to be members of showers long since dissipated beyond recognition, or too weak to be recognized as a shower. However, an exception will occur every so often: a hard, solid particle will enter our atmosphere moving at high velocity. Ranging in size from the smallest piece of grit on up, these meteors have orbits which clearly differ from shower meteors. They are

Fig. 10.8. Several dozen meteors of the spectacular Leonid meteor shower of November 1966 can be seen radiating from the head of Leo, the Lion. Two meteors appear at almost exactly the radiant point. Since they were heading almost directly towards the camera, they look like bright dots. This 3.5-minute exposure on Tri-X film was taken with an f/3.5 lens. Spectacular Leonid showers occur every 33 years, the period of the parent comet.

more like the asteroids or minor planets which have more clearly circular orbits than comets.

On extremely rare occasions these rocky-metallic visitors will survive the high-velocity trip through our atmosphere and become *meteorites*. To be easily recognized as such, a meteorite either has to be seen (or heard) falling or else must land in a snow-covered area or where rocks are few and far between. Even to the trained eye, there is little to distinguish meteorites from Earth material, and only a careful laboratory analysis will be definitive. Many known meteorites have a high iron content – 90 per cent or more – but most look like ordinary rocks and will almost certainly remain undiscovered as a result.

All the meteorites, as well as the much larger asteroids, presumably once existed together as one smallish planet with an orbit lying between those of Mars and Jupiter, but at some time early in the history of this planet it broke apart into billions of pieces, possibly the result of too-rapid cooling. The smaller pieces of this destroyed planet we can see only after they drop to the ground; the larger ones, many irregularly shaped, remain as asteroids.

A freshly fallen meteorite can feel either warm or cold to the touch depending upon its size, its velocity through the atmosphere, and how long it has rested on Earth. Many fall unnoticed and undiscovered since most of the Earth's surface is water, and much of the dry land is not heavily populated. But there have been spectacular meteorite falls where thousands of people have witnessed the passage of the meteor through the sky and the actual fall to ground was seen or heard. So far no person is known to have been killed by a meteorite fall; however, a few have been injured, and some animals have met their final fate astronomically.

Two of the most famous meteorite falls, one prehistoric, one relatively recent, produced large craters, one in northern Arizona and one in Siberia. The first left a lunar-like crater nearly 1.5 kilometers across, and is believed to have been the result of a single solid mass weighing many tonnes which collided with our planet about 20 000 years ago. The other, the Tunguska Fall, devastated hundreds of acres of Siberian Forest on June 30, 1908. A man some 80 kilometers away was blown off his front porch where he was sitting at the time, and much wildlife was instantly vaporized. What is unusual about the Tunguska meteorite is that no meteorite mass or masses have ever been found. Consequently, many believe that the object was a small comet composed largely of ice, which was large enough to survive the trip through the atmosphere but melted away before anyone could inspect the site. However, a competing and intriguing theory is that the meteorite was made of antimatter, material which is known to exist and which has been studied in large atomic accelerators. Its most energetic property is that if brought in contact with normal matter it instantly produces a violent nuclear reaction converting all its mass and an equal part of the ordinary material to energy, in accordance with the $E = mc^2$ formula; 0.03 grams of antimatter (one-tenth the weight of an aspirin tablet) is all that is needed to produce the energy equivalent of almost 1000 tonnes of TNT, the approximate size of the Tunguska explosion. It is lucky for us that antimatter is rare in our corner of space.

It would be almost unbearably exciting to cut into a meteorite and find a perfectly preserved fossil of a primitive insect or fish – or animal! Well, that may have already happened except that the fossil was of a microbe . . . and unfortunately there are reasons to believe that the meteorite had lain around on our planet long enough to be contaminated by Earth-bugs. Another meteorite, cut in two for further study, revealed the clear outlines of a machine screw inside. What was its origin? The mystery was not difficult to solve: the screw had exactly 1.0 threads per millimeter, typical of the hardware used by Soviet space engineers. A piece of a *sputnik* had returned home.

Because meteorites have spent much of their life in interplanetary space, exposed to cosmic rays, they all show trace amounts of radioactivity when they reach the Earth. If laboratory analysis of the type of radioactivity can be made after the fall, much valuable information can be deduced about the age and past history of the meteorite. We now know from such dating techniques that the oldest meteorites were formed at about the time the Solar System was formed, nearly 5 billion years ago.

Would you like some free samples? To get some, you should place a wooden bucket or aluminum pot underneath a rain spout and then, after a rainfall, drag a magnet along the bottom of your rain collector. Some of the iron grit that you pick up this way comes from polluting factories and mills, but much comes from above, mini-samples of asteroids and comets, tiny pieces that the atmosphere slowed down completely and which have been drifting earthward for hours, days or weeks. How do you tell which came from outer space? Only an expert can guess, but a week or so after a major meteor shower (see Table 9.1), the number of particles arriving each day at the Earth's surface goes up noticeably. You might try sampling the old rain barrel beginning tomorrow.

Fig. 11.1. First efforts at astrophotography, with stars appearing less round and the Orion Nebula. Taken with a Pentax camera mounted on an 8-inch Celestron telescope.

Chapter 11: Buy half a telescope

Our pictures have been developed or printed. We have inspected them with a magnifying glass and seen what the motion of the Earth has done to our star images. In the short exposures, the stars are dots and the background stands in sharp contrast. Not too many stars show, but the principal ones are clearly visible.

In the long exposures, the stars trail and register as curved lines instead of dots on the film. When (as in Chapter 7), a 10-second and a 10-minute exposure are combined on the same picture with a pause between them, then a dot (indicating the start of the exposure) is followed by a space where there is no image (while we shielded the lens). Then follows an arc, the length of which depends on the duration of the exposure. The background tends to appear a little washed out. We noted that in pictures taken while we aimed at the North Star, the trails are curved and of varying lengths. The ones nearest the Pole (in the center of the picture) are the shortest; the lines near the edges of the film are longer. This could be compared to the circular distance traveled by a small star ornament near the middle of the hubcap of a wheel on a car and the much longer arc traveled by the valve stem, further out on the wheel. Even during a mere quarter-turn, the former seems to move only a few centimeters, while the latter covered 10 or 15 times the distance. Finally, in the shots taken more nearly overhead, all the star trails are nearly straight. Even though the exposures may have been of the same duration as those we took of the North Pole, the lines are much longer. The nearer to the Celestial Equator they are, the longer and straighter they get. Here again, the valve on an automobile tire does not travel around quite as far as the tread on the outermost and highest point of the rubber tire.

From your photographs and notes, you can determine for yourself how long the ideal exposure for your lens and chosen film should be. If you want to shoot the principal constellations, you will see how long an exposure can be before the stars lose their dot-like appearance and start to turn into lines. The longer the exposure the more starlight is collected. In star-trail photographs, this light is spread out rather than being concentrated in one spot. In all cases we should refresh our memory as we look at the pictures by going back to our records and establishing a relation between the individual exposures and their appearance on the film.

It is time to experiment further: on subsequent nights, we should stop down the aperture ring by one stop from the widest open setting with which we started. Then by two stops and so on. As we stop down, let us increase our exposure. Numbers on the aperture or 'diaphragm' ring are normally marked from the widest open setting (usually f/1.4 or f/2) downwards as follows: f/1.4, f/2, f/2.8, f/4, f/5.6, f/8, f/11, f/16. At these settings there is usually a 'click stop' which you can feel as you turn the ring controlling the aperture. Sometimes there is another click stop between settings which represents 'half a stop'.

When you experiment with any f/stops smaller than your widest open setting, it is very important to remember this principal rule: with every change in number setting (not necessarily corresponding to the click stops), which reduces the size of the lens opening by one stop, the light admitted into the camera is reduced by 50%. Remember that f/2.0 chokes the size of the aperture down to half the f/1.4 opening. For this reason, the length of the exposure must be doubled to permit the same amount of light to enter and to reach the film.

The following explains the concept in general terms and the numbers given can be immediately applied to our work. A 5-second exposure at f/2 should be increased to 10 seconds at f/2.8. A 10-second exposure at f/2.8 should be increased to 20 at f/5.6 and to 40 seconds at f/8. The 40-second exposure at f/8 must be increased to 80 seconds at f/11, etc. Such long exposure times result in ever longer trails. Why, you may wonder, should one try f/2 aperture settings when the camera lens is capable of f/1.4? Well, there is an advantage and once again it is a trade-off. If one reduces the full aperture of any lens by one or two stops, the sharpness of the photograph is greatly enhanced at the expense of a longer exposure time. An f/1.4 lens will thus produce superior pictures at f/2 just as an f/2 lens will make crisper images at the f/2.8 setting. It should be remembered it is not the f/2 setting that is better than the f/1.4 setting, but rather the f/2 setting on the f/1.4 lens. The faster lens will still be the superior one and cost more money as a result. In many instances, the f/1.4

setting is needed, and it is always possible to go back to it if one has the high-quality fast lens. So while one can stop down a f/1.4 lens to f/2 or f/2.8, it is never possible to 'open up' an f/2.8 lens.

We have already proved that the longer the exposure, the more chance there is of the stars turning from dots to short lines. Trial and planning will let you combine these variable options into optimum results. Do not think that long-exposure pictures of star trails are useless, especially when taken with lenses wide open. Repeated long exposures permit 'trolling for meteors' where the chances of catching one are increased with every extra minute the lens is held open.

Still, it is quite certain that you will not be satisfied until you are able to photograph the easily recognizable pictures of stars, which are seen in books or magazines, where every star is almost always of the round dot, non-trailing variety. There is no special magic to taking these pictures: the added element which makes these images possible is corrective camera motion. Cameras with which long-exposure photographs are taken must be made to turn very slowly on an axis parallel to the one on which our Earth turns. If the camera is mounted so that it will turn once in 24 hours or approximately half a turn (180 degrees) per night, star trailing in longer exposures will disappear.

The turning motion of the camera, of course, must be opposite to that of the Earth so that the respective revolutions nullify each other. Don't let this seemingly complex arrangement throw you. All that is needed to achieve the compensating effect and to enable you to keep a camera aimed at one place in the sky for a long time is a clock motor which makes one revolution in 24 hours, exactly half the speed at which the hour hand turns on regular clocks. Such geared-down motors are readily available and, in fact, they form the heart of every electric clock drive mechanism which is built into the mount of better quality telescopes.

Because the motion of the Earth becomes a hindrance in telescopic visual observing also, causing stars to drift continuously into and out of the field of view, such electric drives have become standard equipment on better instruments. Motor drives are also offered as optional accessories to be added to older, fixed telescope systems.

There is a relatively inexpensive method which will allow us to construct a simple 'equatorial drive system' for ourselves. It will 'stop the Earth' and allow us to take astronomical photographs of rare beauty as well as great scientific usefulness. Let us call it the STELAS device (Stop The Earth, Lock All Stars), because the heavens will quite literally be locked in place for the entire duration of our exposures. The system is easy to use and with a 35-millimeter camera and standard lens permits easy star-tracking for periods of up to 15 minutes (see Fig. 11.2).

The heart of the device is a motor which makes one revolution in 24 hours. A camera 'ball-head' is attached to the drive axis which will permit our aiming the camera at any chosen point in the sky. To support the combined weight of camera and lens (slightly heavier and 'longer' telelenses may thus be used) two 'thrustbearings' in the form of small pulleys are introduced. Together with the drive axis they provide a sturdy three-point support.

Construction is of plywood throughout with the motor and other hardware components available from sources listed in the Resources section or from hardware supply stores.

Fig. 11.2. *Stop The Earth, Lock All Stars (STELAS)*.

BUY HALF A TELESCOPE   81

- **STELASCOPE ALIGNER**
  1/4" DIAMETER DRINKING STRAW 6" LONG
- **EQUATORIAL DRIVE MOTOR**
  1/24 REVOLUTION PER HOUR
- **MOTOR SUPPORT BLOCK**
  3/4" PLYWOOD 2 1/4" X 3"
- **MAIN SUPPORT BLOCK**
  3/4" PLYWOOD 4" X 3"
- **TEE NUT**
  1/4" RECESSED 3/8" FROM TOP SURFACE
- **ATTACHMENT TO TRIPOD**
  1/4" TRIPOD SCREW INSERTED HERE
- **THRUST (SUPPORT) BEARINGS**
  SCREEN DOOR ROLLERS (PAIR)
- **CONNECTOR-PLATE**
  GEAR BLANK WITH 1.250 BORE
- **CAMERA BALL HEAD**
  DOT LINE CORP. ITEM, DL-0609

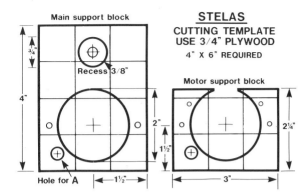

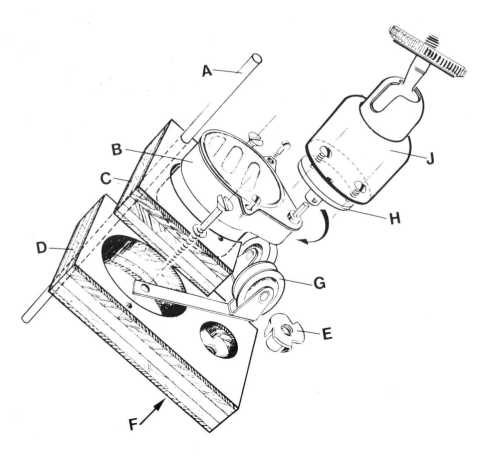

The STELASCOPE used to align the system with the axis of our Earth by sighting-in the North Star is a plain paper drinking straw. When mounted on a tripod and properly polar-aligned, the STELAS system with any standard 35-millimeter camera becomes an important astronomical tool. When used in combination with a STEBLICOM or PROBLICOM (See Chapter 15) meaningful astronomical work can be performed even by beginners from city locations. Transported to dark-sky environments STELAS, when operated via a 12-volt automobile battery, puts the user squarely on the threshold of discovery.

Let us restate our priorities: don't buy a telescope yet. Instead, buy *half* a telescope: that is, the mount *with* a clock drive and start by attaching your camera to that.

Telescope mounts come in several different configurations and usually include a support system. All serve the same purpose though; to keep telescopes or cameras aimed at a selected point in the heavens and to compensate for the apparent motion of the sky. If you were to watch a camera which is mounted on a motor-driven base, with the telelens pointing at a star overhead at 9 o'clock in the evening, you would note that at 3 o'clock the next morning the lens would still be aimed in the direction of the same star which would now be low on the western horizon.

When one contemplates the fact that this same star is at the very moment straight overhead for an observer one-quarter of the world away (6 hours, or 90 degrees farther west) the rate at which our planet turns can be understood even better. If one were aboard a jet flying west slightly faster than the speed of sound, one could follow one's star and keep it permanently in sight, always on high. Just like Alice in Wonderland, one would be going faster and faster just to stay in the same place.

In terms of the heavens, our lens barrel, while describing a 90-degree arc, would actually be standing still also. A little bug which lights on it some warm summer's night thus enters a state of spatial immobility while the world underneath it spins inexorably on.

Buying a motor-driven mount at the outset has many advantages. By investing in the drive mechanism at the start, your camera, when properly attached and aligned, becomes a very powerful astronomical tool. No longer need exposures be held to short durations to prevent stars from trailing. Suddenly the long-exposure, deep-sky photograph becomes an easy reality. With a modest electric drive, the 5-minute exposure of any part of the Milky Way on high-speed black-and-white or color film, through a 50-millimeter lens set at f/2.8, will produce a slide of startling beauty. It is just about impossible *not* to succeed on the very first few tries. If the polar alignment instructions which come with motorized telescopes are carefully followed, results can almost be guaranteed (Fig. 9.1).

You cannot imagine the beauty and splendor that await your patient lens and your unbelieving eyes whose retinas could never collect light as film can. When color slide-film is used, the threshold to the Universe will blaze forth on your projection screen in unsuspected, multicolored glory. Yes, there is color there: dull red stars and blue ones; fierce white diamonds and others with a greenish cast.

By following the fixed stars with the slowly moving camera locked onto them, you can capture light which is completely invisible to the naked eye. With the unaided eye you may see approximately 2000 stars out of the 6000 or so 'naked-eye stars' which can be visually observed in the sky surrounding our Earth. Cameras will bring you hundreds of thousands – even millions more. There are only about 20 brightest stars of first magnitude. Next, there are about 3000 stars ranging from second through fifth magnitude. In the fifth to sixth magnitude category, another 3000 stars are visible under dark skies without optical aid.

A casual survey of the night sky with binoculars will quickly establish that there are many more faint stars to be seen than bright ones. The fainter the stars get, the more of them there are. With a standard camera and a 135-millimeter telelens plus fast film, stars down to tenth or eleventh magnitude lie within the grasp of long exposures. Perfect polar alignment is needed when the camera is left open for long periods and when lenses of high focal length are used. There are nearly one million stars in the range from zeroth to eleventh magnitude.

At or near the ninth magnitude, our own 50-millimeter photographic gateway to the sky, 100 000 stars await us, not to mention the galaxies, nebulae and clusters which can also be recorded. There is so much to photograph that you may want to make two exposures of the areas you shoot: a short shot to pinpoint the main stars, which usually make up the constellations after which the regions of the sky are named, and a second long one to take you to the limit of the capabilities of your film, your camera and the location from which you take your pictures.

You will soon learn how long you can expose before the sky background turns from sharp high contrast to overexposed greys because of 'sky fog' caused by an overall 'glow' in the sky. Light pollution from cities is the main culprit here.

If you do not already own one, a star atlas becomes a sound investment at this point. You will notice that the stars shown in *Norton's Star Atlas*, a classic handbook first published in England in 1910, range only from first (or brighter) magnitude down to approximately magnitude 6. *The Tirion Sky Atlas 2000.0* shows a total of 45 300 stars down to eighth magnitude. Neither book can even approach the reproductive powers of your first, few photographs.

Your short exposures will most likely be the ones which will allow you to find your bearings as you compare your pictures to star charts or atlases. When I promise fabulous pictures if you point your camera at 'any part of the Milky Way' it is because this faint, mystical band of countless stars is our own Galaxy seen from within. The Solar System of which we are a part is merely an infinitesimally small speck in this galactic system comprised of thousands of millions of other suns. We know that our galaxy is a spiral disc: thick in the middle, thin around the edges. Our own Sun is not at the center of our Galaxy as originally believed, but way out, nearer the edge of our galactic saucer.

How can one find one's way in this celestial maze? How is it possible to put order into such a cosmic labyrinth? Does one lose one's way forever or is it possible to go back to one selected point in the heavens, perhaps to photograph it again and again and watch a variable star change from a tiny speck to a large fireball and back again?

The ancients named the constellations and referred to them in terms of shapes which were familiar and helped them find their bearings. Omar Khayyam spoke of 'that inverted bowl which they call the sky'. Of course we know that there are really two inverted half bowls which comprise the entire sky around our Earth (Fig. 11.3). This is the celestial sphere around us, a cage of vertical rings and horizontal hoops. It becomes a convenient gridwork for measurements. Vertical lines are called circles of Right Ascension. Horizontal lines are called circles of declination. The Earth rotates counterclockwise as seen from the North Celestial Pole within this cage (see Fig. 13.1).

The largest horizontal hoop forms the Celestial Equator which lies above the Equator of our Earth. This is where the two inverted sky-bowls touch. Any point in space about us can be established, either in degrees of declination angle above or below this zero line, or in hours and minutes of Right Ascension on its circumference. It is merely a question of 'how high above or below the celestial horizon' or 'how far around on the 24-hour clock of Right Ascension'. Just as the prime Greenwich Meridian on Earth remains at zero longitude, so the prime celestial meridian forms the zero from which all Right Ascension measurements start.

There are no hard and fast rules concerning the size of the celestial sphere itself, since it is merely a concept. Better to think boldly and in cosmic terms. Perhaps it is best to visualize a balloon which can be blown up more and more as our ability increases to see and feel our way farther and ever farther out. How far is infinity?

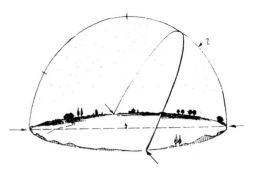

Fig. 11.3. 'That inverted bowl which they call the sky.'

Fig. 12.1. The Earth as photographed by the astronauts on board Apollo 16. The USA and other parts of North America can be clearly seen.

CHAPTER 12

The Earth in motion

When you think about it, buying 'half a telescope', as advocated in Chapter 11, permits the accomplishment of a rather overwhelming feat: it compensates for the spinning of the world and provides a platform which remains rigidly oriented relative to the stars. That this Earth-platform travels around the Sun in a circle with a radius of 149 597 870 kilometers is of less consequence and produces relatively inconspicuous effects. The Sun and its family of planets and other miscellaneous debris also move in other ways: among the stars at 20 kilometers per second, around the nucleus of our pinwheel-shaped Galaxy at 250 kilometers per second, and outward in the general expansion of the Universe at who knows what speed. These motions generate only negligible movements of most of the stars in the course of a lifetime.

When they are counted up, the motions of the Earth are seven in number. Above, we have described five of these; the two others are the periodic disturbances typical of any spinning top: precession and nutation, two subtle aberrations of our planet's spin-axis, one of which produces a small but readily noticeable effect, namely precession.

Let us now consider the more important Earth movements beginning with the most obvious one: rotation.

The spinning world

Look again at Fig. 13.1, the sketch showing the globe with a camera and tripod mounted on it and aimed at the North Celestial Pole. In your mind, picture what the Earth's rotation would do to the appearance of the sky as seen from a point on Earth. To an Earth-based observer, infinitesimally small on the scale of Fig. 13.1, only half of the sky would be visible at one time. Imagine if you will a large, flat piece of paper placed edge-on at a tangent to the globe, touching only at the point where the sub-microscopic observer stands; this plane defines the horizon and divides the sky into two, above and below. With the plane of paper still firmly attached to the world, see what happens when the Earth rotates on its axis. (If you have handy a spinnable globe, you might actually pin a piece of stiff paper or cardboard to your latitude and longitude and demonstrate what follows.) As you slowly turn the globe, the piece of paper turns itself around the Earth's spin-axis, taking its slice out of the surrounding sky in continually changing ways. Remembering that the micro-observer can only see what is above the plane (in a direction away from the earth), notice that unless you live exactly on the Equator, there will be a piece of the sky that never becomes visible, even after a full rotation. A polar bear located right at the North Pole (or a penguin at the South Pole) would never see more than half the sky, the same half no matter how long the Earth rotates. The closer to the Equator you live, the larger fraction of the sky you get to see.

Sighting along the plane of the paper, you can visualize how the Earth's rotation causes the rising and setting of the stars and Sun. Again, the observers at the Poles have special viewpoints: risings and settings never occur. The astronomical bodies viewed from either Pole move in circular paths parallel to the horizon; the North or South Celestial Pole is directly overhead. At the Equator, an observer sees everything rise and set with a movement precisely at right angles to the plane of the horizon. At inbetween latitudes – for example, consider half-way up from the Equator at 45°N – Polaris, the North Star, located less than a degree from the North Celestial Pole, describes a small circle around the Pole once a day and remains forever visible half-way up from the horizon to the zenith, the imaginary point directly over the observer's head where he is standing. Other stars nearby, for instance in Ursa Minor (Polaris marks the end of the 'dipper' handle), likewise move in circles about the Celestial Pole as do the stars in the Ursa Major. At a 45 degree latitude, all these stars remain continuously above the horizon. The word used to categorize these stars is *circumpolar*: stars which always are above the horizon.

Before we consider what happens in the rest of the sky, let us again define the celestial coordinates, essential as they will be in locating objects described in past and future chapters. In Chapter 11 we introduced the prime meridian of the sky, the 'hoop' that ran from the North to the South Celestial Pole and through the Greenwich of the sky, called the *Vernal Equinox*, a location also known as the 'first point' of Aries. This is where the Sun is located at the first instant of spring (northern hemisphere) or fall (southern hemisphere). Any star, planet, or nebula positioned on that imaginary arc has a *Right Ascension* of zero. Right Ascension is to the sky as longitude is to the Earth: it tells us how far east of the Vernal Equinox a star is. Right Ascension is usually expressed in units of time, and it always *increases eastwards*. One complete circle around the sky is divided into 24 hours. Since 360 degrees

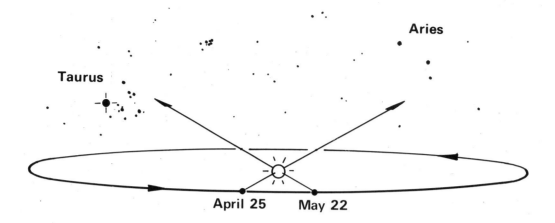

Fig. 12.2. If we could see stars behind the Sun, Aries and Taurus would be the constellations located far beyond the Sun during the northern spring, as this figure illustrates. In October and November when the Earth is on the other side of its orbit, these constellations are high in the sky at midnight.

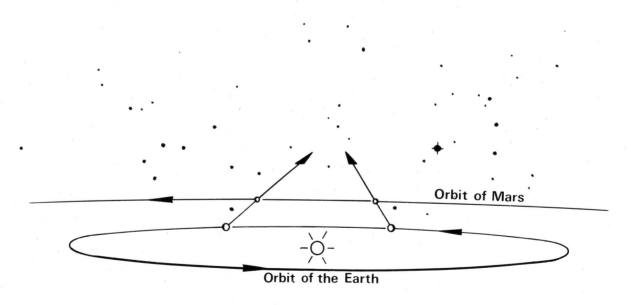

Fig. 12.3 Moving faster on an inner track, the Earth passes by Mars every 2 years and 50 days on average. As it does, Mars appears to move backwards in the sky as this diagram illustrates.

defines a full circle, 1 hour = 15 degrees. Fig. 13.1 will help explain these concepts.

On some clear, warm night, lie on your back, your head toward the north, feet toward the south, and east to your left, and imagine yourself looking up at the sky dome marked with these Right Ascension circles. All pass through the North and South Celestial Poles (only one of which will be visible unless you live on the Equator), spreading apart until they meet the *Celestial Equator*, the great circle located exactly halfway between the poles. If stamina permitted it, you could, in your imagination, watch these hour circles pass by, one after the other, all night long (and all day long, too), rising up from the eastern horizon, crossing the center line of the sky, and setting in the west. That center line in the sky, by the way, is called the *celestial meridian*; it's the great circle that runs through the two Celestial Poles and the observer's zenith directly overhead. It also intersects the horizon at the north and south points.

In a very real sense, the sky acts like a giant celestial clock with the centrally located meridian marking off the hours of Right Ascension as they march by. The Right Ascension of all celestial objects on the meridian is, in fact, what we call star time, or more properly, *sidereal time*. If you want to know what the sidereal time is at this very moment, you have to know the Right Ascensions of the stars on the meridian, or vice versa. It is a most useful concept to keep in mind: *sidereal time equals the Right Ascension of the meridian*.

As you probably have already guessed, an hour of star time is not quite the same length of time as an hour of normal time (solar time). It gains 10 seconds every hour, so that a sidereal day is completed in 23 hours and 56 minutes of solar time. The reason for the difference has to do with the Earth's orbital revolution about the Sun, something we will discuss in the next section.

The other celestial coordinate – the latitude of the sky – is called *declination* and can be simply defined: it runs in a series of parallel circles from zero degrees at the Celestial Equator to +90 degrees at the North Celestial Pole and −90 degrees at the South Celestial Pole.

Spend a few minutes with a good star map, one that has a grid of Right Ascension and declination marked clearly on it. One useful exercise is to identify the location of zero hours Right Ascension, zero degrees declination (0^h R.A., $0°$ Dec.) relative to the stars. This special location, the first point of Aries, actually now lies in the constellation of Pisces and forms the southeast corner of a small triangle with the stars labelled Lambda and Omega. When we discuss precession later, we shall see that this fundamental point in the sky actually moves, creeping along steadily westward and a little southward at a rate of about one degree every 72 years.

Also, notice on your star map where other prominent stars and constellations are located in this grid of R.A. and Dec. – Orion at 5.5^h R.A., $0°$ Dec.; Leo at 11^h R.A. $+20°$ Dec.; Scorpius at 16.5^h R.A., $-30°$ Dec., and so on.

Returning to the appearance of the sky at Earth latitude 45 degrees North, you can confirm from Fig. 13.1 that the stars north of $+45$ degrees declination are all circumpolar.

Spend some time familiarizing yourself with the celestial sphere, not just memorizing constellations. Spend an hour just watching a bright star rise in the east or set in the west. At what angle to the horizon does it move? Get the feel of how the grid of coordinates criss-crosses the sky. Know especially where the meridian and the Celestial Equator are located and the point where they cross. How high in the sky is this point? How far is it from the zenith? What is the sidereal time? These are valuable concepts and you will quickly grasp them, gaining at the same time a better understanding of the sky.

The annual trip around the Sun

During the year, the Sun moves through different constellations, specifically the 12 constellations of the Zodiac. In mid-March, the Sun crosses the border from Aquarius to Pisces, and by late April moves into Aries, and so on.

However, our days and our entire life are governed by where the Sun is located in our sky, not where it lies with respect to the stars. Most of us awake around sunrise and go to sleep several hours after sunset. Our clock runs on Sun time. All the year, throughout each year, the stars and constellations shift gradually with respect to the Sun in such a way that every night at a given time we are able to see almost exactly one degree more eastward. In fact, the ancient astronomers of Babylon, believing that a year was 360 days long, divided the great circular path of the Sun through the stars into 360 parts, or as we now called these divisions, degrees.

The seasons

The Sun appears to move continuously along one great circle in the sky, the *ecliptic*. When plotted on a flat, rectangular star map, this circle becomes a wavy curve which oscillates from declinations $+23.5$ to -23.5 degrees crossing the Equator at Right

Ascensions of 0 and 12 hours. The crossing points are called *equinoxes*, since when the Sun is there day and night are of equal length; at the ±23.5 degree declinations, the Sun is at one of the *solstices* where it halts momentarily in its northerly or southerly movement. When the Sun arrives at these celestial points, we mark the beginning of spring, summer, fall or winter, depending on which hemisphere we live in. Notice especially that the Sun is at the Vernal, or spring, Equinox on or about March 21 every year. (The exact time varies because of leap years.) Why does the Sun take this celestial roller-coaster ride rather than a straight journey along the Equator? It's because the flat plane defined by the Earth's Equator and the flat plane defined by the Earth's orbit are inclined to each other by an angle of 23.5 degrees. If you lived at an Earth latitude of 23.5 degrees, north or south, once a year on the first day of your summer, either June 21 or December 21 (give or take a day), the Sun would pass right through your zenith because of this tilt. If you lived at a pole, for half the year – your winter – you would remain on the dark side of the Earth continuously. Come the first day of spring the Sun would finally rise to the point where during the day it would slide all the way around the horizon. From then on through spring, the Sun would slowly spiral up to its maximum altitude of 23.5 degrees and then, with the beginning of summer, it would turn around and start the long spiral down.

For reasons having to do with how the Solar System was formed, all the major planets plus the Moon have orbits which are very nearly in the same plane, and therefore we see the Moon and the planets very nearly on the ecliptic at all times. Most of the asteroids and many of the comets likewise stay close to the ecliptic.

The Moon, never more than 6.3 degrees from the ecliptic, moves in its orbit about the Earth once every 27.3 days, but because in the meantime the Earth–Moon System has moved farther along in its larger orbit about the Sun, it takes another 2.2 days for the Moon to line up with the Sun again, adding up to a total of 29.5 days for the completion of all lunar phases. Meanwhile of course, the Earth keeps spinning rapidly on its axis, giving Earth-dwellers the illusion of a Moon rising in the east and setting in the west. Gradually the Moon seems to drift relative to the stars until it moves from the night sky into the day and passes in front of the Sun (or nearly so). The slippage amounts to an average of about 13 degrees a day or a little more than half a degree an hour. It just so happens that the Moon's diameter is also a little more than half a degree, which means that the Moon moves through the stars by one of its own diameters every hour. This is a handy fact to know if you are waiting for the Moon to pass in front of a bright star or planet.

Other earthly motions

With the exception of an axial wobble called *precession*, all the other movements of the Earth produce miniscule effects detectable only with carefully made measurements. As noted, these include the lesser wobble known as *nutation*; the motion of the Solar System as a whole through the stars around us; the revolution of the Earth, Sun, and neighboring stars around the center of our spiral Galaxy; and the movement of our entire Galaxy outward from some as yet undetermined point in the Universe.

Precession is a sizeable Earth wobble, large enough to produce a noticed effect, but it takes a little patience to observe. At the present time, the Earth's axis points very nearly to second-magnitude Polaris, but at other times it did, and one day it again will, point to other stars – Thuban, Vega, Alpha Cephei, to name a few. At the same time the two directions defined by the intersection of the planes of the Earth's orbit and the Earth's Equator, the Equinoxes, will sweep slowly around the sky along the ecliptic.

The time for a single round-trip oscillation is long: 26 000 years. The practical result is that the direction of the starting point of Right Ascension and declination, 0 hours and 0 degrees – the Vernal Equinox – is constantly moving. In Babylonian times, this point was in Taurus ('The gleaming Bull opens the year with golden horns', Virgil wrote much later in 2450 BC); by the time of Christ it had slid into Aries; and today it is in Pisces. The beginning of 'the age of Aquarius', when the Equinox moves into that constellation, is still more than 600 years in the future, according to the modern location of the boundary between Pisces and Aquarius.

To the professional astronomer, precession requires making a continual updating of the Right Ascensions and declinations of celestial objects at intervals depending on the accuracy required. Every 10 or 20 years will usually suffice if he has a small telescope; the professional using a 4-meter telescope sometimes needs to 'precess' his coordinates on a monthly or even weekly basis. Atlases are usually updated every 50 years.

So the Earth moves, spinning, circling, spiraling through space, accompanying the Sun wherever it goes, and sharing in the Grand Plan, the expansion of the Universe. The motions are a thing of beauty, although they can also be a darned nuisance. But the astrophotographer is fortunate and only has to compensate for one of the Earth's movements – rotation – to get nice, star-dot images. The 'half a telescope' of Chapter 11 does exactly that. But before this celestial camera mount can be used, it must be correctly oriented with respect to the Earth's axis.

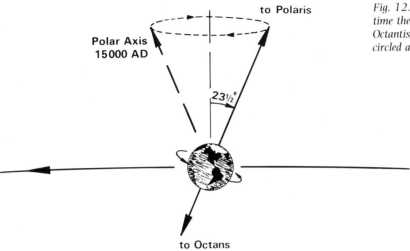

Fig. 12.4. The Earth as viewed from the Sun on March 21. At the present time the Earth's axis points towards Polaris in the north and Sigma Octantis in the south, but 13 000 years from now the axis will have circled around to points in the sky 47 degrees away.

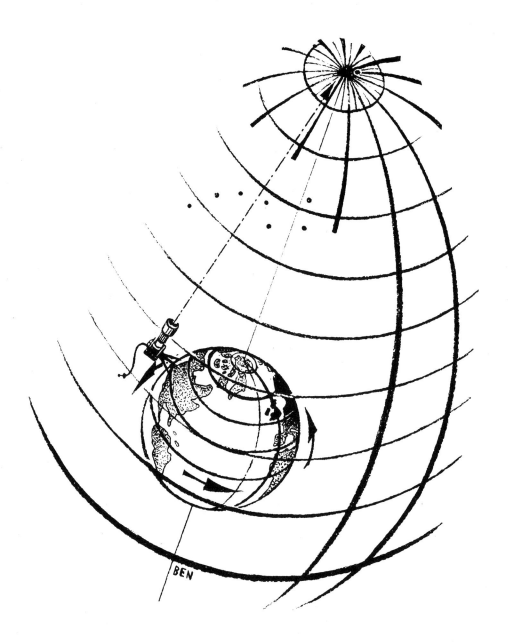

Fig. 13.1. The celestial sphere is a cage of rings and hoops surrounding our Earth. It creates a convenient gridwork for measurements. The Earth turns counterclockwise within this sphere as viewed from the North Celestial Pole.

CHAPTER 13

Two down, three across

On Earth we pinpoint exact locations anywhere on the globe by using a ball-shaped grid system involving longitudes and latitudes. Navigators of old and pilots of the twentieth century are using the same globe-girdling horizontal and vertical lines for reference.

At the risk of overstating the importance of the celestial grid system, let us relate it to the making of our own photographic star atlas. If you think in terms of maps of our world, the half-moon-shaped meridians connecting the North to the South Pole are the lines of longitude. The circular demarcations, smallest near the Poles, largest at the Equator, are the lines of latitude. As we have seen, only a very slight modification of these concepts is necessary to locate stars or other objects in the heavens.

Let us imagine the celestial sphere in the form of an enveloping, transparent balloon surrounding our globe with the stars, the planets, the Moon, the Sun and any other celestial objects painted on it. On this sphere, the degrees corresponding to latitude on our Earth are expressed in terms of degrees of declination (Dec.). The lateral degrees of longitude become hours and minutes of Right Ascension (R.A.). We have already found that for the 24 hours of longitude, the zero line passes through Greenwich, England. Back when Britain decreed that this line, which actually cuts through the telescope pier in the Royal Observatory Building, would be the place from which all points on the Equator would be measured, the Royal Navy ruled the waves and with it much of the sea commerce of the world. Maritime traffic needed exact charts and so the seemingly arbitrary but highly practical system came into being. To this day, the framework, formulated long ago, remains the basis not only for all terrestrial but also for celestial navigation. In terms of longitude, Greenwich is at zero, but being quite a ways north of the Equator, its latitude is 51 degrees north. Any point north of the Equator, celestial or terrestrial, is expressed in degrees north or plus (+). Whatever lies 'below the waist' of the Equator is said to be x degrees south or minus (−).

Only a very slight modification of terrestrial concepts of navigation is needed to make these apply to the heavens. As long as we keep referring back to Polaris, our task will remain simple. Let us move away from our planet for a moment and look down on the Northern Hemisphere from the North Star. Earth's North Pole would be in the center of a circular shape, which could be said to represent a clock. If we position ourselves so that the Greenwich Meridian is at the top of the circular shape, then we can read the time clockwise around the Equator in 24 hourly increments. The 12 hours of daytime plus the 12 hours of darkness would represent a one-day revolution of the Earth. Half of our globe would in effect, be in darkness, the other half in the light, illuminated by the Sun.

In a way we are quite spoiled. Recent photographs taken from orbiting spacecraft and satellites have given us a new perspective and a much better understanding of our Earth. Watching satellite weather reports on television allows us to actually place ourselves above our world, a feat which earlier generations could only dream about.

There is one principal difference between using the 360-degree grid system on Earth and in space. It lies in the fact that on our planet, there is an east and a west, a north and a south, but hardly any 'up and down' or 'in and out'. On Earth, all points measured are on one and the same surface, an equal distance from the center of the world. Even the peak of Mount Everest is only a tiny fraction further out than the deepest trench in the oceans. These are the outermost and innermost places on the surface of the Earth. The difference then lies in the third dimension in space. Declinations indicate north-south, and Right Ascensions denote east–west measurements, respectively. The coordination of R.A. and Dec. is what will help us find any place in the heavens at any time, again and again. The third dimension, that of 'in and out', of 'how deep in space', is one which need not be discussed in detail because it will not help us find our way any better in the celestial sphere. Still, to place ourselves into the perspective of space, it is necessary to speak briefly about the staggering aspects of 'depth', in terms of 'how far out'. It all depends on where you set your boundaries, close in or far away. Let us start by referring to the angular distance of one degree. This unit, used for measuring the two-dimensional spans of the sky, changes as soon as a third dimension, depth, is introduced.

If you think in terms of one degree on the Earth's Equator, then this one degree of angle constitutes about 110 kilometers. (Originally, a kilometer was defined as one-ten thousandth of the distance from the Equator to the North Pole along the Earth's surface.) As soon as you raise your sights nearer to the celestial sphere, say a mere 160 000 kilometers above the surface of our planet, you're speaking about lateral spreads of 2790 kilometers per degree. The area subtended by the one-degree angle continues to widen until it reaches a spread of distances which are beyond comprehension. What is happening, in effect, is that as the imaginary spheres surrounding our Earth increase in size over and over again, we reach the uncharted depth of deep space.

To try to relate such vastness to the human scale, let us extend our arm to its full length and hold our forefinger to cover the half-degree wide disc of the Moon. By checking, we will find that a finger-width covers the Moon twice, which makes the human digit a convenient one degree measuring-scale. The Moon has a diameter of 3500 kilometers, and so we can easily tell that at lunar distance, the width of our finger equals about 7000 kilometers. The angular distance between the two pointer stars in Ursa Major is 5 degrees. Dubhe and Merak are an average of 85 light years away from us. At such a distance, the width of our hand, or the 5 degrees between the prime stars in Ursa Major, becomes 7.4 light years. Applying the rule of thumb and forefinger to the 2 million light year-distant Andromeda Nebula, which is 5 degrees wide, the width becomes 175 000 light years.

Where does infinity begin?

If we select celestial areas to photograph, we have 41 253 square degrees of sky from which to choose. That is the number of such one degree × one degree squares which surround our globe. Let us see how many photographs would be needed to shoot this cosmic panorama if we used a standard 50-millimeter or 55-millimeter lens. The field which such optics cover is approximately 24 × 36 degrees in size. This means one photographic 35-millimeter frame will record more than 800 square degrees. If nearly 42 000 square degrees make up the heavens around our Earth, then theoretically about 50 slide shots would cover the entire area and provide us with a complete sky photo atlas of our own, in black and white or in color.

Yet only one half of the celestial globe is visible from any one point on Earth at a time. One's place on the planet determines which part of the sky can be seen. For North Americans, Europeans and North Asians, the Northern Celestial Hemisphere is the principal area. To cover half the inverted bowl, 25 shots, in principle, is all one would have to take. Working fairly fast and under a sky evenly clear and dark in all directions, it would be possible to perform such a feat in one long night on one roll of film.

But let us consider the layers of turbulent atmosphere which surround our globe with obscuring moisture and dust. It becomes clear that in aiming the camera straight up, perpendicular to the obscuring layers, the light lost from a distant star would be held to a minimum. Light from objects which must be photographed at low altitudes has to travel laterally through many miles of obscuring atmosphere. For Northern Hemisphere observers, this will be of concern only for stars far in the south. All other objects will present themselves overhead in gradual succession. Over a period of 12 months, they will seem to rise in the east, 'culminate' on the meridian, and set in the west. The best time to photograph them is while they are near the zenith, that is as near to straight above as possible. If you plan to collect your atlas photo records during the periods of almost totally dark nights nearest the New Moon, you will be out and photomapping about 12 times a year, weather permitting.

One overhead 'band' of pictures at a time will fill the need. Center the first one on the Pole Star. Shoot others so they overlap a little and work your way south along the meridian, ending up near the southern terrestrial horizon. This kind of survey can become a most important part of your first year of photography and give you your own reference material to which you can compare future photographs. The camera should be mounted so that it shoots a 'horizontal' picture as you aim it facing Polaris. Then each series of photographs being more than 2 hours (or 30 degrees) wide will produce a complete record; 12 monthly photomapping sessions multiplied by 2-hour wide pictures equals 24 hours. There will even be extra portions on the sides of each frame to create some overlap. The number of pictures needed to take a vertical panorama of sequential shots can be easily determined: we start with a shot centering on the North Celestial Pole, then work our way south. Let us assume we are on latitude +45 degrees (north) half-way up on the Earth between the Equator and our North Pole. This means that the Celestial Equator, which parallels our own, is 45 degrees above our southern horizon. Thus, we have 45 degrees between the Celestial Pole and Celestial Equator plus 45 degrees from the Celestial Equator to the Earth's southern horizon, which equals 90 degrees of declination in the sky to shoot. The 45-degree portion which lies between Polaris and the northern terrestrial horizon can be disregarded because in 6 months time any objects which lie beyond and beneath the Pole will be straight overhead and in a much better position for photography.

We have established that a

50-millimeter or 55-millimeter lens takes in a field approximately 2 hours (30 degrees) wide by 24 degrees high. We use the width portion for the lateral hour-spread of our maps. This leaves us the vertical dimension of each photo with which to create our composite slice of the heavens. Twenty-four divides into 90 degrees five times, with some degrees of overlap to spare. So, five horizontal frames 12 times a year will give us a complete sky atlas from the vantage point of $+45$ degrees on our globe. (In fact, since Polaris remains virtually motionless throughout the year, the shot aimed at the North Celestial Pole need not be repeated every time.) Also, you can complete the photo atlas in fewer sessions by shooting at the meridian both before and after midnight and taking advantage of the motion of our Earth during each night. If we lived on the Earth Equator, we could theoretically take the greatest number of pictures because one can see both the Northern and Southern Hemispheres from such an equatorial point with the North Celestial Pole visible on the northern horizon and the South Celestial Pole on the southern horizon. In Equador, then, to record a 180-degree slice of the heavens, we would need eight frames, each 24 degrees high. Again we would have a little overlap, a little to spare.

Here is what you do with a motorized equatorial mount equipped with 'wheels' showing hours of Right Ascension and degrees of declination. Align your camera drive with the axis of our planet. Then center a known bright star in the viewfinder and set your Right Ascension wheel and your declination indicator to its coordinates. These can be found in charts or atlases. You are now completely lined up and ready to find any other objects in the heavens from the coordinates given in sky maps or in books. You will have precisely synchronized your motorized base to the accurate 24-hour Right Ascension clock of the heavens. Should you want to find any other object hereafter, all that is needed will be for you to swing the camera so that the Right Ascension wheel points at the selected hour. The declination disc should be set on the chosen angle. Zap – you have hitched your astrophotography system to the stars!

It is inevitable that you will want to have an easy method to familiarize yourself with the night sky to find specific starfields or targets. The purchase of a fine star atlas, such as the *Tirion Atlas 2000.0*, can be an invaluable aid. When combined with inexpensive 'STARFRAMES™' which you can create yourself out of simple wire coat-hangers, you can become acquainted with stars and constellations simply and quickly.

The starframe method is completely new and was first developed by me in the book *Starwatch*. The concept is based on tracing bright principal stars from suitably scaled graphics in an atlas and then holding the transparent tracings up to the sky and lining them up with their celestial counterparts. Ideally, you should hold starframes against the heavens at the time of culmination of constellations, i.e. when the respective stars are at their highest positions in the nightsky.

The way to achieve easiest results is for the observer to be 'polar-aligned' with the north–south celestial axis on which our planet revolves. This can be achieved by lying down, head to the north, feet pointing south (for observers in the northern hemisphere).

Here are a few pointers on how to prepare a starframe. Bend a wire coat-hanger as illustrated in Fig. 13.2. Then stretch some transparent kitchen-wrap on the frame as shown. Such wrap adheres to itself and can be made as taut as a transparent window if the completed starframe is held in a warm oven for a minute or two.

To copy the brightest stars from an atlas, use white typewriter correcting fluid; this is fast-drying and can be seen easily by a red-masked flashlight without spoiling the night vision of your dark-adapted eyes.

Much better yet are stardots copied with luminous phosphorescent paint (or ink) available in hobby shops. You need only the smallest bottle of such water-reducible non-toxic paint. Use a toothpick to flow the stardots on to the film while the entire starframe lies on the atlas for copying. Trace the brightest stars with larger dots and the fainter ones with smaller marks. Indicate special points of interest with a cross or circle. You can combine luminescent paint for the stars with brush-written notes using typewriter correcting fluid. Allow all inks to dry.

Now you 'charge', or activate, the phosphorescent paint by placing the entire starframe in a lightproof cardboard box with a white light in it. Do not expose your eyes to such glare, so that you stay fully adapted to the dark.

If you lie down in a polar-aligned position, with your head to the north and feet pointing south, your starframes will always have north at the top.

The angles at which the frames should be held depends on the position of stars or their constellations in the sky and the latitude of the position of the observer on Earth. For example, viewing from northern terrestrial latitudes of approximately $+40°$, a starframe for Polaris and Ursa Minor would be held at $+90°$, up and behind the head of the starframe observer as shown in Fig. 13.3.

To view Orion or Virgo from similar latitudes, wireframes would be held at an angle of $+2$ celestial degrees, see Fig. 13.4. To view southern constellations such as Scorpius or Sagittarius from northern-hemisphere locations, one points the starframe down in the

94 THE CAMBRIDGE ASTRONOMY GUIDE

Fig. 13.2.

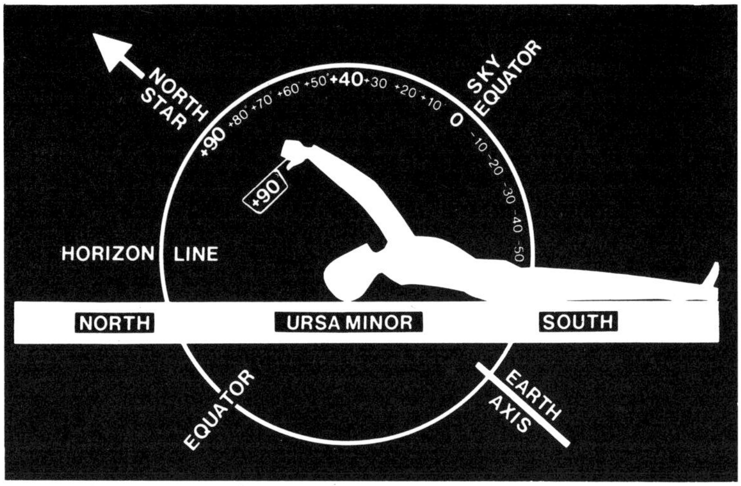

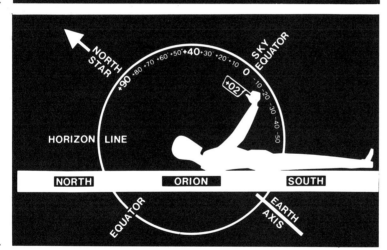

Fig. 13.3.

Fig. 13.4.

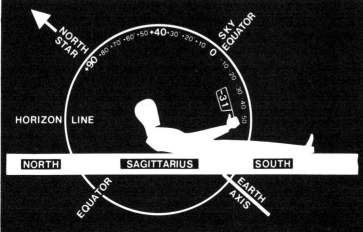

Fig. 13.5.

direction of −32 or −31 degrees as illustrated in Fig. 13.5.

Subscription to a good monthly astronomical publication will update readers all year round, telling which starfields are in the sky for best viewing, and their positions in Right Ascension and declination (up, down and across). Such magazines also point out special objects to be found in each constellation and in the sky in general. The finding of special targets, newly discovered comets or novae is greatly simplified using this system.

Starframes are ideal guides for astrophotography. The area outlined by the wire matches exactly the 'field of view' of a 35-millimeter SLR camera with a standard 50-millimeter or 55-millimeter lens. Held approximately 18 inches from the observer's eyes, a starfield of 30 degrees vertically by 2 R.A. hours across (near the celestial equator) is delineated.

It is important for anyone just beginning to do astrophotography to frame his photographs exactly as the wire frame suggests. The angle at which the starframe is being held to encompass specific stars should be noted and copied when the camera is aimed before opening the shutter. In this manner your photographs will show the precise area which the star atlas depicts, and you will be able to identify countless new objects in each photograph.

By photographing matching pairs of pictures on subsequent nights, weeks months or years, you can become a 'blinking astronomer' and place yourself on the threshold of discovery.

Fig. 13.6. (overleaf, pp. 96–7) *Horsehead Nebula, IC 434, in Orion.*

Fig. 14.1. Comet Austin 1982, as photographed in August of that year, had a skinny tail composed almost entirely of ionized gases. The broad tails seen in other comets (see Fig. 8.10, for example) are made up of grains of dust and metallic particles.

CHAPTER 14

Scientific fallout: the contributions of the amateur

In 1980 there were just over 4000 professional astronomers in the world; there are 41 253 square degrees in the sky: 10 square degrees per scientist. If amateur astronomers did not exist, thorough patroling of the night-time skies would be difficult, even if each professional were to agree to watch a different 10-square-degree piece of sky on a continuing basis.

About half the professional astronomers in the world are theoretically inclined and probably know less about cameras and telescopes than you, the reader of this book. About half the observational astronomers do their research at radio frequencies, in the infrared, or via rockets and satellites. Perhaps half the Earth is cloudy at any given time and of course less than half the world is in total darkness at this or any other moment. Finally, if you consider the remaining available professional observers, not all are able to get out and view the sky on a given night because of other interests or commitments. What are we left with? We find that the scanning of the skies is left to a dozen or so professionals, most of whom are looking for bizarre things such as gamma-ray sources and distant quasars.

Consider the scenario of that night of the nova, August 28–29, 1975. The Northern Hemisphere nights were delightfully warm and astronomically inviting, and the summer Milky Way passed overhead shortly after darkness. Observing conditions were ideal in many places. While a handful of professionals were manning giant telescopes to analyse the light from faint, distant stars, nebulae and galaxies, the light arriving from one particular faint, distant star located in Cygnus (the Northern Cross) was rapidly increasing in intensity. The Earth's rotation had brought central Canada and the United States under the Northern Cross at the time this exploding star had reached naked-eye brightness. When Cygnus was over Hawaii, the star had reached fifth magnitude, and when it reached magnitude 3 the star's light shone down on Asia.

An amateur astronomer in Japan, Kentaro Osada, was the first person to recognize that a 'new' naked-eye star had appeared in Cygnus, and as the Earth continued to turn, more and more independent discoveries were made, largely by amateurs, first throughout Asia, then Europe and finally in the United States and Canada.

The point of this story, told from a different perspective in Chapter 11, is that there were very few professionals among the several hundred independent discoverers of Nova Cygni. The same can usually be said about first sightings of most bright novae and comets – and many faint ones, too. Professionals have to plan their observing programs months in advance, and then they devote their time to the detailed study of specific objects in the sky. They rely heavily on the amateur to tell them where the new and the temporary phenomena occur.

In astronomy the amateur can make truly important contributions to the scientific world. Astronomy provides a tremendous variety of specimens – stars, planets, galaxies, nebulae – none of which can be manipulated in the laboratory or in the field. Only by watching patiently can we expect to see the unexpected, the changes which provide us with additional clues concerning the origin, physical nature, and eventual fate of literally billions of individual celestial objects.

Enter the amateur

If there is one quality which an amateur must have to guarantee making a significant astronomical discovery, it is persistence. As we have emphasized over and over in this book, the simplest of cameras can be turned into a powerful observing tool, but it can survey nothing if it remains stashed away in a closet or bureau drawer. If you really want to discover something, take your camera outdoors at twilight whenever the sky is clear and the Moon not too bright, set it up, and when dark enough, start shooting.

If you have the talent to make little gadgets, do so and make the job as automatic as possible. Timing motors are one of the amateur's best friends because they provide by simple, adjustable cams and microswitches the necessary signals to open and close shutters, advance film, and even change the direction in which the camera points. (See Chapter 19 for further thoughts and ideas.) Do not feel that you have to sit up all night while your

Fig. 14.2. (overleaf, pp. 100–1) Comets do not always come with magnificent tails. This photograph of Comet Kohler (left page, right of center) taken by the two authors in September 1977 shows it shortly after discovery. A typical long-period comet, it should be coming around to our neighborhood again in 100 000 years, give or take a few millenia.

camera works; you should get a good night's sleep so that you can awake early, develop the film (or deliver it to the local photolab), and inspect all the frames as soon after shooting as possible. Later, even while your camera has been put to work again, you can pore over the previous night's output at leisure.

A frustration which you will encounter is that when you do see something of interest on your film, you will have to wait until the next clear evening or morning to rephotograph the area to confirm its existence – and all but the most spectacular 'discoveries' must be confirmed, if not by you, then by a fellow sky-watcher with whom you have previously entered into a discovery-sharing pact. For this reason a good scheme is to rig your automatic picture-taker to make two exposures, one right after the other. Then at least photographic flaws can be identified.

The reasons why you should confirm your discovery before informing the world is obvious: photographs *can* lie, either through defects in the film put there by the manufacturer or by you or whoever in the darkroom (and you can never be sure which), or by ghost images produced usually by a bright star or planet either in or just out of the field of view. Remember that if you do get fooled by a flaw and report it, then the next time you will have problems convincing the astronomical world that you mean it when you really do discover something. Furthermore, you will have taken up the valuable time of some, we presume, hard-working professional to whom you first reported your imagined discovery.

This section started with the statement that persistence is the most important quality for a productive amateur to have. But you have to have luck, too. Do not get discouraged if the 'comet of the century' was discovered half a degree off the edge of the picture you took two nights before. With persistence your time will come.

How to discover a comet – or a nova

Once you have become expert at taking pictures of the stars and have accumulated a few dozen stunning photographs of the Milky Way towards your sky atlas, or perhaps a definitive collection of constellation portraits, chances are you will start itching to get active in a special field of astronomy. Perhaps variable stars. Good! There are plenty for all and observations are always valuable. But perhaps you had your mind set on discovering something, on being the first person on Earth – or in the Universe – to see a new star or comet or asteroid, and getting your name attached to whatever it is. In this section, we will tell you where in the sky discoveries are made, who has been making them, and exactly what to look for.

First, consider comets (Figs 14.1 and 14.2). In the 5-year period ending December 31, 1982, a total of 37 comets were discovered and verified. (Seven others were reported but not verified.) Four of the 37 turned out to be comets that had passed by this way before but whose orbits were poorly known or had been drastically altered by the gravitational attraction of a planet or two, or never came sufficiently close to the Earth to get bright enough to be seen again. One comet, after being discovered and verified, went out of sight before there was time to derive an orbit accurate enough for recovery in some future epoch. In addition, several 'Sun-grazers' were seen briefly, comets which were recorded by solar telescopes as their orbits carried them through the Sun's corona.

Of the 32 remaining newly discovered and well-tracked comets, 17 were fainter than magnitude 15, and one was given as magnitude 13.5–14; all were discovered photographically by professionals. Fourteen were brighter than magnitude 12, and 13 of these 14 were discovered by amateurs. (The fourteenth was discovered by two Soviet astronomers in Uzbekistan; whether they were amateurs or professionals is not known). The last statistic is worth repeating: 13, possibly all, of the 14 comets discovered in the 5-year period 1978–1982 and brighter than magnitude 12 were discovered by non-professional astronomers. It is worth noting emphatically that a twelfth-magnitude star can be seen with a telescope as small as 3 inches in diameter, or with 12×80 binoculars. Three of the comets were visible to the naked eye at time of discovery – or would have been except for their proximity to the Sun.

Some more interesting facts: five of the 'amateur comets' were discovered by one man, William Bradfield who lives near Perth, Australia, and who systematically sweeps the sky with a 6-inch refractor set up on a home-made wooden mount. Three more were found by Rolf Meier of Ottawa, Canada, with a 56-power, 16-inch reflector. Other discoverers include an Englishman, a Japanese, a South African, a New Zealander, another Australian, and one solitary observer from the United States, Don Machholz, who used a 10-inch reflector.

Comets can be capricious. Five amateur comets (and one professional) were discovered in the 40 nights beginning September 1, 1978; no amateur comets were found during the almost 18 months between December 25, 1980, and June 18, 1982. Bradfield reported one discovery made on Christmas Day, 1979; on December 18 of the following year, he discovered another comet at a position in the sky less than a degree away from the discovery position of the first comet; and on Christmas Day, 1980, Roy Panther, the Englishman, found another comet.

Where did they find all these comets? The answer: nearly all within 75 degrees of the Sun. (Only one was farther than 90 degrees from the Sun.) There was only a slight affinity of the comets to the ecliptic; the average angle above or below this zodiacal circle was 36 degrees. Seven were found in the evening sky; seven in the pre-dawn sky.

How many will be seen again in our lifetime? Only one of the amateur comets, comet Haneda–Campos, which orbits the Sun once every 5.97 years. The others have periods ranging from a few hundred years to 'forever'. However, 10 of the 18 professional comets have short periods (less than 20 years) showing that almost all the reasonably bright short-period comets have been discovered.

This last statement can also be made about asteroids. To discover an asteroid, an amateur needs either a large camera–telescope or a lot of luck. On very rare occasions, an asteroid will pass within 1 or 2 million kilometers or so of the Earth and then become bright enough to be 'captured' on a photograph taken with modest equipment. Remember that in 1938 the asteroid Hermes passed by the Earth at a distance of 800 000 kilometers and in 1983 Comet IRAS–Araki–Alcock missed us by less than 5 million kilometers. Normally both would have had a greatest brightness in the vicinity of fifteenth to twentieth magnitude. Who will be the one who discovers the next 'once-in-a-lifetime' close passerby? Probably the amateur who conscientiously patrols the skies looking for the new things out there. We hope it will be a reader (or writer) of this book.

Now some words about novae. In the same 5-year period, 1978–1982, 16 novae were discovered. Much of the story of their discovery is like that of comets: 10 were brighter than magnitude 12 at time of discovery and all but one – or perhaps two – were found by amateurs. One of these was a recurrent nova, WZ Sagittae, found by a professional; the other was picked up by Peter Collins, a computer expert at the Whipple Observatory in Arizona who 'moonlights' as a sky-patroler when he has the time. His nova was independently discovered by at least four others.

The remaining eight bright novae were all found by one person, a Japanese amateur named Minoru Honda, although on one occasion the first report came from another astronomer, also a Japanese amateur. On another occasion, an independent discovery was made by Ken Beckmann of the United States, an enthusiastic variable-star observer and AAVSO member. After the initial excitement had calmed down following two of Honda's discoveries, it turned out that the reported novae were actually long-period variable stars, one of them previously known and properly catalogued (V 3876 Sgr). Thus, we are left with six or perhaps seven amateur novae, and eight novae brighter than twelfth magnitude at time of discovery.

Minoru Honda used, at least for some of his discoveries, sky photographs taken with Tri-X film. Yoshiyuki Kuwano, the other Japanese astronomer mentioned above, also used this film. Collins carefully memorized all the stars he could in the Northern Milky Way and made his discovery visually.

The discovery magnitudes of the 10 bright (or suspected) novae ranged from 6–7 for Nova Aquilae 1982 to 10 for V 3876 Sgr. All were located in, or very close to, the Milky Way, with three appearing in Sagittarius, two in Cygnus, and one each in Serpens, Vulpecula, Corona Australis, Aquila, and Sagitta. None of the novae got much brighter after discovery, but Nova Cygni 1978 could be faintly seen with the naked eye at sixth magnitude a few days after it was first reported.

Summarizing the statistics, we find that as far as amateur discoveries are concerned, the average in recent years has been about two new novae and three comets each year. There are strong reasons to believe that these numbers could be considerably larger with more patrolers using the new blinking techniques, especially so with novae. One of the best-known experts on novae, the late Cecilia Payne-Gaposchkin, estimated that approximately two dozen novae occur every year brighter than magnitude 9.0. The astronomer H.C. Arp at Mount Wilson–Las Campanas Observatories discovered 30 novae in a one-year survey of the Andromeda Galaxy, thought to be almost a twin of our own spiral system. To be sure, if 30 novae occurred each year in our Galaxy, we wouldn't be able to see them all since interstellar matter in the form of crystals, grains, gas and dust blocks our view of much of the Galaxy. It's a big place, too. But it would be reasonable to expect a half dozen discoveries per year, perhaps eight or ten in good years.

Speaking of other galaxies, can an amateur discover novae there? Not unless he has a large camera–telescope. The brightest nova that Arp found in the Andromeda Galaxy had a magnitude of 15.7. There are nearer galaxies, but except for the southern Magellanic Clouds, they are small and produce very few novae. The Large Magellanic Cloud, or the LMC as it is more often called by the professionals, will have two or three novae per year on average, and their peak magnitude will be around 11, easily in reach of 35-millimeter cameras with telephoto lenses and fine-grain film. A nice target if you live far enough south.

Three or possibly four novae have been found in the vicinity of globular clusters, those fuzzy patches in the sky, often with Messier numbers, containing up to a half million stars each. Nova hunters should pay special attention to these objects, many of which lie along the Milky Way and in the regions favored by the more usual novae.

Hunting for supernovae in other galaxies is also a worthwhile project and recently, one amateur, Robert O. Evans of Australia, has piled up an impressive number of discoveries (three within 41 days!) using a home-made 10-inch Newtonian and a fantastic memory for star fields, rarely using maps. We should

Fig. 14.3. The discovery photograph of Nova Muscae 1983 taken by this author with 2415 film, an f/1.4 lens and with a 55-millimeter focal length. At magnitude 7.1, the nova shows up clearly in this 5-minute exposure. Compare with Fig. 6.2.

add, however, that Gregg Thompson and James Bryan, Jr, have prepared an atlas of star maps near several hundred nearby galaxies, complete with magnitudes of selected comparison stars near each object. This atlas certainly should be in the library of every amateur – and professional – interested in discovering supernovae.

In a good-sized galaxy like our own or the Andromeda Galaxy, a supernova goes off once every hundred years or so. There are clusters of galaxies which make it possible to photograph on one piece of 35-millimeter film dozens of galaxies – not all as large as our own to be sure, but so numerous that one should expect supernovae to be seen reasonably often. One of the biggest and closest is the Virgo cluster, centered on M87, an immense flattened sphere of stars considerably larger and more populous than our own Milky Way Galaxy. Containing roughly 2500 galaxies, the Virgo cluster produces no more than one or two supernovae per year, the brightest averaging about twelfth magnitude. Other clusters of galaxies farther away and usually less populous generally do less well in their annual production of supernovae.

We save for the end of this section a report on the discovery of Nova Muscae 1983. It was made on January 18, 1983, by this writer, from the front porch of his home in Viña del Mar, Chile. The nova was found at magnitude 7.1 on photographs of the Southern Milky Way taken with fine-grain film (Kodak 2415) and a standard 35-millimeter camera with a 55-millimeter focal length, f/1.4 lens (Fig. 14.3). With the 'blinking' technique described in Chapter 15, there was no need to memorize star positions or to compare laboriously each star on the new photograph with its image (or lack of same) on a comparison print. Nova Muscae literally blinked out its presence amongst the teeming thousands of other stars that crowd pictures of this brilliant region of the Milky Way just south of the spectacular Southern Cross. To make absolutely sure that the image of the newcomer was real and not a ghost or emulsion effect, the region was photographed again the following night (with considerable difficulty owing to patchy, thick fog blowing in from off the nearby ocean). On the morning following the second night, the following telex message was sent:

TELEX 710-320-6842 (ANSWERBACK ASTROGRAM CAM) CENTRAL BUREAU FOR ASTRON. CAMBRIDGE MASS. POSSIBLE NOVA JAN 18.1420 RA 11H 49.5M DEC MINUS 66D 55M EPOCH 1950 MV 7.2 JAN 19.15 MV 7.3. DISCOVERED WITH PROBLICOM. 3.7M EAST 23M SOUTH OF MU MUSCAE. WM. LILLER, VIÑA DEL MAR, CHILE.

Within minutes of the arrival of this message, Dr Brian Marsden, the Director of the Central Bureau for Astronomical Telegrams, retransmitted the message to several observatories in the southern hemisphere and to the European headquarters of the only orbiting astronomical telescope, the International Ultraviolet Explorer. Several hours later, Marsden received confirmation from M.D. Overbeek in South Africa that there was indeed something new at that position in the sky, and subsequently spectrographic observations revealed without a doubt that the object was a nova. On January 20, the official announcement was made on Circular No. 3771 of the International Astronomical Union.

Discoveries and what to do about them

We will now assume that there on one of your pictures you have a smudge, fuzzy image, or a diffuse trail which was not there the last time you photographed that part of the sky. Furthermore, you have examined the image with a high-powered magnifier or microscope and are convinced that it is a real image and not a piece of dandruff or a scratch or a stain on the emulsion. It is, you feel certain, a new comet. If you are working with someone else, call him up and ask for confirmation; otherwise, you must wait until the next opportunity and photograph it again or look for it with a telescope. The only exception to this Golden Rule is when the object is so bright and so positively real in appearance that there can be no doubt. Experience is a great teacher. Practice 'discovering' comets by taking pictures of known comets when they are faint and see what they look like on your films (see also Fig. 14.2).

Because a comet will move on average a degree or so a day, one piece of information you will need to know when you find your comet is in which direction it is heading and how fast. Therefore, confirmation photographs taken 24 hours later should show your comet clearly and a degree or so away from its original position. If your discovery was made in time to get a second picture within a few hours of the first, the intervening motion may not be conspicuous, but with careful inspection it should be detectable.

If by now you are *absolutely certain* that you have discovered a comet, get to the telephone and contact the nearest observatory, university astronomy department, or planetarium. If you are lucky, you will end up talking with a professional observer with whom you can discuss what you have found and how you found it. (*Remember*: many theoretical astronomers and radioastronomers will know less about the appearance of the sky than you.) In fact, if you plan to get into the business of searching for comets or whatever, you should post the telephone number of your nearest useful (and hopefully friendly) astronomer just below the numbers of your local police department, fire department, etc.

Talking to a professional does more than help you decide if you have really discovered something. He may also provide you with the news, disheartening to be sure, that your comet is an old one, that someone else has already discovered it, or that it is merely a comet-shaped nebula not listed in your catalogue.

Also, be considerate with him. If he is the only professional astronomer within miles, he probably hears other amateurs too, and since amateurs outnumber professionals by 10 or 20 to one, he may spend a considerable amount of time each week talking to others like you.

But now you are as sure as you can reasonably be that you have found something that no one else in the whole world knows about. Then do as I did and send a telegram or cable or telex to: Central Bureau for Astronomical Telegrams [TELEX No. 710–320–6842 (Answerback Astrogram Cam)], Smithsonian Astrophysical Observatory, Cambridge, MA 02138, USA. For a comet discovery, you should clearly state the following pieces of information: (1) the location of your object; (2) epoch of coordinates; (3) the time of the observation; (4) a brief description; (5) the brightness; (6) the direction and rate of motion; and (7) your name and address. A typical message might read like this:

DIFFUSE OBJECT NO TAIL FOUND AT RA 6 HOURS 34.2 MINUTES DEC PLUS 14.7 DEGREES EPOCH 1950 AT 9.4 HOURS U.T. AUGUST 23. VISUAL MAGNITUDE 10.5. DIRECTION OF MOTION SSW, 1.3 DEGREES PER DAY. (SIGNED) W. HERSCHEL, 132 MARLYBONE LANE, BATH, ENGLAND.

The more accuracy you can provide the better, particularly with the coordinates and the corresponding time of observation. And **DO NOT** forget to include the epoch, the date to which the coordinates correspond. With this information, the astronomers in the business of calculating orbits can get to work on your celestial 'baby' and predict its future path through the sky and its brightness.

In response to your telegram you may get a reply asking for more information. (Sometimes messages get garbled in transit.) Eventually you will receive word confirming (hopefully not otherwise) your discovery. Some observatories subscribe to the telegram service of the Central Bureau. You should ask your local professional astronomer if his institution does, or where the nearest subscriber might be. Within a day or two, all discoveries are reported on the *Circulars* of the International Astronomical Union. Basically these are nothing more than postcards containing news and information about discoveries and other important astronomical happenings. They are the 'voice' of the Central Bureau for Astronomical Telegrams. Earlier we suggested that you might subscribe, especially if there is no astronomical center nearby that does. One can subscribe, as we have said, for not too much money; they are essential if you are serious about making discoveries. Only with these *Circulars* can you check to see if your new discovery is actually Comet So-and-So discovered a week before in Uzbekistan, Nagasaki, or Lima.

There exist several journals which will tell you about known comets due to pass into our part of the Solar System. The yearly *Handbook* of the *Royal Astronomical Society* of Great Britain, and the *Observers' Handbook* of the *Royal Astronomical Society of Canada* are especially valuable and inexpensive references to have (see p. 172 for addresses). Also magazines like *Sky and Telescope* and *Astronomy* keep you well informed and give a more current review of the cometary happenings in the sky.

If you think you have discovered a nova, the procedure is much the same as for a comet find. Make absolutely sure that the new speck on your film is really a star image. If you have or can borrow a telescope, look at your prospect. Not only does this guard against ghost images or photographic defects being the offending source of your pride, but also you can take immense pleasure in actually *seeing* your new star, even if someone else saw it first. Unlike a new comet, it will be in the same place in the sky the following night.

There is one other possibility: your nova candidate may be a known variable star which in recent months has lurked below your magnitude limit and then during a cloudy stretch blossomed into a star bright enough to record its presence on your film. Remember Honda's two prospects. There are several ways to guard against what could be a sad embarrassment. The AAVSO's *Variable Star Atlas* will have just about every star you can record down to about ninth magnitude; for fainter objects you will have to refer to the official listing of all known variables, a five-volume work called the *General Catalogue of Variable Stars* and now being published in the Soviet Union. Check with your local astronomer to see where the nearest set is. Or after a year's time, you will have accumulated a valuable storehouse of information yourself. Go back over your past sky photos and see if the new star is in fact one that you recorded but happened to miss for some reason. Even after all this, you may end up reporting a new variable, as Honda did; it's nothing to be ashamed about. In fact, in some ways discovering a new variable is more important than discovering a nova. They stay around longer.

Clubs, societies and associations

The most useful amateur group to join is your local one, because the direct help you can give and receive from others is the best. To find out where and when the group next meets may take some effort, but inquiry of your local astronomy department, planetarium or observatory is usually enough to get you started.

The larger amateur societies, some of which are listed in the Appendix, have the advantage of being better informed since they specialize in one or two fields and have input from all over the country or the world.

Probably best known is the one we keep mentioning, the AAVSO, or the American Association of Variable Star Observers. Truly international in scope, the AAVSO disseminates finding charts and comparison star magnitudes to all who join. It continually collects, analyzes and publishes members' observations. The AAVSO is also the publisher of the most valuable atlas of variable stars. A close working relationship has developed between the AAVSO and the professional astronomers of the world because of its important contributions.

Other societies and associations in the United States include the Association of Lunar and Planetary Observers who also circulate bulletins on the brighter comets, the American Meteor Society, and the American Meteoritical Society. A valuable (and inexpensive) bimonthly letter is called *Tonight's Asteroids*, which includes easy-to-use star maps of the brighter or more interesting minor planets as well as bright comets.

To conclude, persistence and luck are the key words to remember, if you have your hopes set on discovering a comet, nova, asteroid or supernova; and while only comets and occasionally asteroids take the name of their discoverer, you can be assured that just being associated with an astronomical discovery will make it all worthwhile.

Circular No. 3869

Central Bureau for Astronomical Telegrams
INTERNATIONAL ASTRONOMICAL UNION

Postal Address: Central Bureau for Astronomical Telegrams
Smithsonian Astrophysical Observatory, Cambridge, MA 02138, U.S.A.

TWX 710-320-6842 ASTROGRAM CAM Telephone 617-864-5758

POSSIBLE NOVAE

W. Liller, Viña del Mar, Chile, reports his discovery of a possible nova in Norma at $\alpha = 16^h09^m51^s.0$, $\delta = -53°11'32''$ (equinox 1950.0). The object was detected at mag 9.4 on Sept. 19.00 UT but must have been fainter than mag 11.5 on Sept. 10.0.

E. P. Aksenov, Sternberg Astronomical Institute, reports that N. E. Kurochkin has discovered a possible nova in Triangulum at $\alpha = 2^h42^m.2$, $\delta = +33°19'$ (equinox 1950.0). The object was observed at $m_{pg} = 15.0$ on Sept. 11.96 UT.

OCCULTATION OF 14 PISCIUM BY (51) NEMAUSA

A preliminary report by D. W. Dunham, Computer Sciences Corporation, indicates that extensive positive and meaningful negative observations of this Sept. 11 event were made throughout the southeastern United States. The northern limit was effectively defined by M. Croom, who observed a $2^s.3$ occultation at Yorktown, VA. The longest reported chord, $13^s.5$, by R. Melvin in central North Carolina, corresponds to a diameter of 160 km for Nemausa. The southernmost positive observation, by E. Seykora and others at Greenville, NC, gave an $8^s.2$ occultation. A few observers reported secondary extinctions, but these are believed to be due to seeing or transparency variations or photoelectric guiding errors.

COMET ČERNIS (1983l)

Ephemeris continuation from the elements on MPC 8139:

1983/84 ET	α_{1950}	δ_{1950}	Δ	r	m_1
Sept.23	$1^h45^m.81$	$-10°46'.1$	2.460	3.372	10.2
Oct. 3	1 26.48	-15 22.7			
13	1 05.90	-19 36.7	2.490	3.410	10.3
23	0 45.54	-23 10.9			
Nov. 2	0 26.77	-25 57.6	2.730	3.459	10.6
12	0 10.64	-27 57.9			
22	23 57.68	-29 19.0	3.109	3.516	10.9
Dec. 2	23 47.94	-30 09.9			
12	23 41.24	-30 38.9	3.541	3.582	11.3
22	23 37.20	-30 52.9			
Jan. 1	23 35.42	-30 57.4	3.961	3.655	11.6

1983 September 23 Brian G. Marsden

Fig. 14.4. A Circular *of the International Astronomical Union chosen not entirely at random. On these cards most of the important discoveries in astronomy are announced.* Circular *No. 3871 appeared several days later reporting that Dr Ruth Peterson observed the spectrum of the Norma candidate and confirmed that it was a nova. Nova Tri was confirmed several days later at the Royal Greenwich Observatory.*

Fig. 15.1. Dr William Liller, mentor and friend, at a costly blink comparator, a precursor of PROBLICOM.

CHAPTER 15

The blinking astronomer

It may be just as well that amateurs seldom get to see the costly equipment used by the professional astronomer. The price tag on almost any accessory found in astrophysical science departments is so astronomical that it is light years beyond the reach of even the affluent amateur.

One such piece of instrumentation is the blink comparator, yet no self-respecting department of astronomy would be without one. The machine is used to examine pairs of photographs of the same region of the sky taken on different dates. If the photographic plates are properly aligned in the comparator, and then presented alternately to the eye in fairly rapid succession, any image that has moved, brightened or faded, appeared or disappeared, immediately attracts attention.

Traditionally, blink comparators are table-mounted devices with oculars resembling microscopes. One sits in a chair and looks into either monocular or binocular eyepieces. Mechanical, optical or electronic systems are used to bring first one picture and then the other into view. If the two pictures are properly aligned, the viewer perceives one stationary sky image although two star photographs are actually blinking on and off alternately at a rate of two or three times a second. As stated above, any change between the two photographs fairly jumps at the observer precisely in the rhythm at which the pictures appear in the eyepiece.

The opportunity for discovery which such a system offers can readily be understood when one thinks of comparing two photographs, each containing thousands of stars, one taken in August, the other in September. It would be virtually impossible to detect any difference between the two or to spot, in the later plate, a tiny newcomer which might herald the blossoming of a nova, or the arrival of an as yet undiscovered comet. Not so with a blink comparator. Every star would remain static except for the steady pulsing of the tiny new light source.

Perhaps there is no accessory more important and potentially valuable to the amateur than such a 'discovery machine', but unless one has access to considerable funds, and highly specialized catalogs, it is nearly impossible even to find the sources for such equipment, let alone purchase it.

There is one simple and inexpensive way to build one's own system, a project that is far easier than one might anticipate. The very need for simplicity in construction rules out traditional eyepiece devices where prisms are used to split beams. Stereoscopic versions are bound to be even more complex and totally beyond the capabilities of the weekend hobbyist. Instead of the eye-straining method of squinting through a microscope-type device and viewing pictures through an ocular, the projected images of simple slide projectors should be considered. Usually the pictures are large and easily seen and can be viewed by more than one person at a time. Through the combination of two projectors and by superimposing a projected picture with a later date onto the identical projected image of an earlier one, a very simple new method for blink comparison suggests itself.

Even the crudest experiments with two matching slides prove the feasibility of the system: by simply blocking out the beams alternatively with one's hands, the blinking effect can be simulated and tested. All that remains to be done to create a working blink comparator is to motorize the alternate covering and uncovering of the projection lenses. A motor of 100–120 revolutions per minute, mounted between the lenses, turning a semicircular cardboard shutter to interrupt the beams alternately, will perform the task reliably (see Fig. 15.2). To bring the lenses closer to each other it is best if the two projectors are stacked one on top of the other (no trays or magazines are needed because only one pair of slides at a time is shown). With leveling knobs on the two projectors, the pair of projected images are matched so that they overlap *precisely*. When the principal stars in the two fields coincide accurately on the screen, you are ready to turn on the motor. Sit back, relax and watch the pictures for anything that jumps, flashes or blinks. Pairs of color slides or black-and-white negatives can be studied in such a comparator. The realism of a dark blue night sky will add to the pleasure of viewing. Ordinary black-and-white negatives can be inserted in cardboard mounts and inspected immediately after development, often permitting blink comparison within an hour of photography.

It is important that the pictures to be compared be taken through the same optical system, whether camera or telescope. Film type, exposure times, etc.,

Fig. 15.2. (a) *A PROjection BLIink COMparator (PROBLICOM) consists of any two matching slide projectors with an occulting shutter mounted in front of the lenses. A motor allows the light beams to be alternatively interrupted presenting two different slide images to the viewer(s) in rapid succession.* (b) *Template for fabricating a PROBLICOM out of a piece of plywood $48 \times 12 \times \frac{3}{4}$ inches.* (c) *Exploded drawing showing hardware requirements and assembly of a PROBLICOM in easy steps.* (d) *Schematic drawing of PROBLICOM platforms for two projectors.*

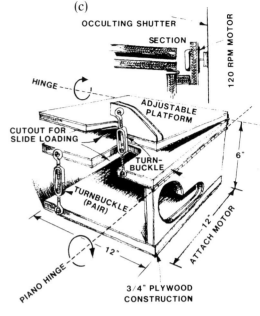
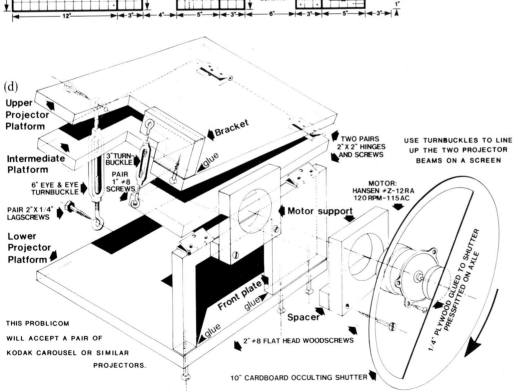

should also be similar whether the pictures are taken two nights, a month or a year apart. Finally, it helps if the field of view is precisely the same in both exposures. By setting the camera to the same Right Ascension and declination, and photographing identical regions on different nights, it is possible to begin a scientifically valuable photographic patrol program with minimal effort or equipment. Try taking six or seven regular 50-millimeter lens exposures and capture most of the Milky Way at least once every month in season. Then compare these photos with the ones taken 4 weeks earlier.

Load up with high-speed film, set the camera at the 'wide open' setting and expose for 20 seconds with a tripod mount or for 8 minutes with an equatorial drive. If you possibly can, duplicate the set on another night (or immediately after the first) in case a photographic defect suggests a discovery where there is only a flaw in the film emulsion. Slides are usually sharp enough to let you find easily the planet Uranus at sixth magnitude in its orbit or seek out charted asteroids by merely photographing the general region in which they are predicted to be at a certain time. Variable stars burst forth like flashing beacons and there's always the element of hopeful anticipation that comes from looking for the unpredictable, for the discovery not yet made by anyone.

Perhaps you want to try a smaller area, say, in Sagittarius or Cygnus where many novae are discovered, and expose a carefully programed roll of film with a 135-millimeter lens taking in a smaller field or a 300-millimeter lens capturing an even smaller rectangle.

It is safe to say that you will be virtually alone in these endeavors. Professionals do not blink 35-millimeter negatives or color slides. They work with large glass plates in highly localized areas. Their costly equipment may not even accept a standard slide or cardboard mount.

With the projection method, many people watching a screen simultaneously will be able to share in the excitement of the search. Among members of an astronomy club, several may have similar slide projectors, which makes the construction of a blink comparator an easy joint effort. If the 'over and under' rotation shirt board shutter does not appeal to you, try the 'side by side' double-barreled shotgun approach and mount an automobile windshield wiper with a large occulting blade in front.

As yet there is no projection blink comparator on the market. This really means that there is a lot to be done with such an easy-to-build combination. If used by enough people, they will bring larger areas of the sky under closer surveillance and yield important results. Much as the AAVSO has aided professional astronomers for almost 100 years, so regular patrol programs conducted by amateurs will inevitably produce important spin-offs. In this new way, everyone can become immediately and directly involved in the rewarding pursuit of astronomical study or even research.

Because there are no projection blink comparators commercially available as yet, there has never been a mention of an organized blinking program for amateurs. Blinking is an endeavor serious amateurs either never knew about or dismissed as belonging to the realm of the professional only. What a pity on the one hand, but on the other, what an opportunity. Given no more than a camera with a standard-type lens or a telelens, a drive mechanism to keep the system tracking equatorially and a pair of slide projectors (one of which may be borrowed or shared), *anyone* can set out to do useful and potentially important work. One need not even be particularly familiar with the stars or the sky.

There is another way to construct a discovery machine. It is the simplest type of blink comparator and utilizes only two small switchable nightlights and two identical magnifiers which should be approximately '8 power'. (They should magnify our slides by a factor of eight.) I call it the STEBLICOM (Fig. 15.3).

It bears mentioning that the successful hobbyist will think of countless alternative ways of constructing such a simple device. A sturdy cardboard box can be employed instead of the plywood version suggested. Frosted glass may be substituted for the acrylic slide support platform. Glue or masking tape may be used to assemble the entire package where care has been taken that the basic components are safely in compliance with electrical codes when they are purchased in a hardware store.

The principle governing the use of the STEreo BLInk COMparator is the same as outlined at the beginning of this chapter. The methods for use are also identical. First you align the pair of transparencies by sliding them into a position where they appear to form one stereoscopic picture when viewed with both eyes through the magnifiers while both lamps are on. Persons possessing very acute observational gifts or 'stereopsis', a special kind of stereoscopic vision, will be able to detect any difference between slides at this preliminary stage. Others may need to turn one of the two lamps under the slides on and off in order for the eyes to be guided to the moving planet in the pair of slides or the diffuse image of some comet which may herald an important discovery.

The introduction of motorized or solid-state blinking devices into the circuit of one of the two nightlights offers

112 THE CAMBRIDGE ASTRONOMY GUIDE

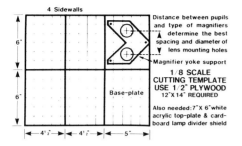

4 Sidewalls

Distance between pupils and type of magnifiers determine the best spacing and diameter of lens mounting holes

Magnifier yoke support

Base-plate

1/8 SCALE CUTTING TEMPLATE USE 1/2" PLYWOOD 12"×14" REQUIRED

Also needed: 7"×6" white acrylic top-plate & cardboard lamp divider shield

- **A ▸ TWO 7 WATT NIGHT LIGHTS** WITH ON/OFF SWITCHES.
- **B ▸ THREE-WAY PLUG** EPOXY GLUE TO BASE-PLATE.
- **C ▸ 3/8" PLASTIC TUBING** OR USE FELT-PEN CAPS.
- **D ▸ LAMP DIVIDER SHIELD** 4"×5" LIGHTPROOF CARDBOARD.
- **E ▸ TWO 6-8 POWER MAGNIFIERS** OR COMPARATORS WITHOUT RETICLES.
- **F ▸ MAGNIFIER YOKE SUPPORT** MEASURE DISTANCE BETWEEN PUPILS.
- **G ▸ 7"×6" WHITE ACRYLIC TOP** OR USE FROSTED PLEXIGLASS.
- **H ▸ VENTILATION HOLES** CAN BE TRIANGULAR CUTOUTS.
- **I ▸ (3) WOODEN SPACER DOWELS** TO PERMIT ALIGNMENT OF SLIDES.

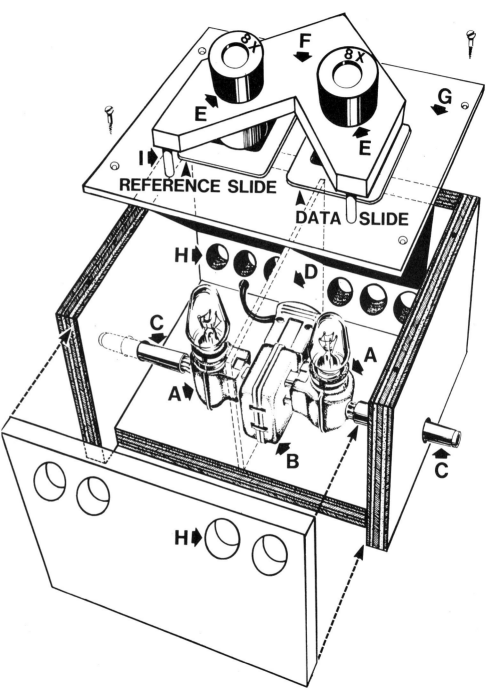

Fig. 15.3. This is a home-built STEBLICOM (STEreo BLInk COMparator). Any sturdy box will serve as a housing for two low-wattage lamps and as a support for a sheet of frosted acrylic. You need switch on only one of the lamps and can do so with any simple circuit-breaker. Only UI- or BS-approved components should be used, so that the apparatus is safe.

additional opportunities for modification and upgrading of the STEBLICOM with commensurate improvements in the ease of using the device.

The learning process concerning celestial motions will be greatly enhanced because rather than study the movement of the Moon or of the planets, of asteroids or comets, by means of charts or graphs in books, one can take sequential pictures over a period of several nights and by 'blinking' these, one can actually observe the paths of the wandering objects. Being able to trace their course and relating their previous positions to the present ones, one may even venture to make predictions of where things might be on following nights. Phrases such as 'the retrograde motion of the planet Mars' assume new meaning. This traveling backwards of The Red Planet, which gave so much cause for puzzlement to the ancients, can be recorded, studied and therefore understood by anyone. We can suddenly identify with the shepherds in Galilee, who lay in the fields night after night viewing and observing the motion of the heavens and movements within it, without understanding or even the hope of grasping the cause of the magnificent but strange behavior of the celestial objects. Ancient Ptolemy's passion to put order into the heavenly motions can be better appreciated and his brilliant – but incorrect – solutions evaluated in light of the gifted astronomers who followed him.

Even when we look up to the stars night after night (and weather may stop us on occasions) it is difficult to detect real movement there. Only the Moon seems to be going places. Lunar movement and, of course, the daily celestial happenings which bring us sunrises and sunsets, are the only signs of motion of which we are strikingly aware.

Blinking one's own photographs can bring home with almost staggering impact the supremely inspiring forces causing the motions of heavenly bodies. Instead of just reading about wandering planets, they can be seen and recorded on a day-to-day basis as they travel through constellations which become gradually more familiar to new observers as they hitch their cameras first to the larger and eventually to the lesser celestial travelers which journey with us about the Sun.

Even though astronomical publications will, in their monthly issues, print maps or photographs showing where to find Uranus or even faint and distant Neptune on specific dates, the fact remains that when you go out on the appointed night and look exactly at the given position, lo and behold not only do your binoculars reveal a tiny spot – which might be the object of your search

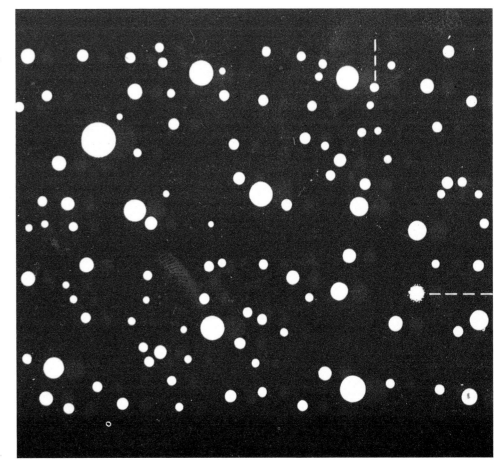

Fig. 15.4. A 'blinking' effect can be obtained by flipping this starfield photo back and forth rapidly over that below (on page 115). You should be able to find a nova, a comet and a variable star.

– but two others exactly like it, causing you to wonder, quite validly, just which one is the planet and which are the stars. Aim your telelens at the designated area a few nights in a row – the longer you wait between pictures, the bigger the 'jump' caused by motion will be. Almost immediately your doubts will be removed once and for all. More yet, with your new knowledge in mind you can then walk out into the night, find your object again and if you had a fair number of shots to study, predict where your planet will be tomorrow or even next week.

A rewarding pursuit, which brings much satisfaction and new understanding, is the charting of the position of the planets together with dates in the sky atlas, which, by this time, you may well have bought. *Norton's Star Atlas*, the old standby, may have been your choice, but you may want to add to it some more recent or even more detailed atlas of the heavens which not only gives you large-scale maps but also an acetate overlay grid which makes plotting and charting a veritable delight.

Mark the position of your objects in pencil on the charts, write the date next to them and perhaps the number of the film and the frame of the picture showing them. This will allow you to retrieve valuable information from your log at future dates. You have been keeping a log haven't you?

You don't have to shoot too many pictures or photograph too often; a long-range well-planned program will serve you best. The first picture entered will pinpoint your planet in the heavens; the second gives you direction of travel and approximate rate. Three shots thus recorded on your charts may allow much more: three shots can show an apparent slowing down, speeding up or curving around, which may cause you to delve more deeply into planetary motion to find out exactly what is going on out there and why.

If you read in the newspapers or astronomical publications about a newly discovered comet, you know, of course, that you will have a head start with your system for capturing the heavenly visitor and identifying it long before it has become a naked-eye object and long after it has faded from view for most other visual observers. Call your local observatory if the publication does not give the necessary information – ask for the coordinates of the celestial visitor. Even though this information changes from night to night as the comet hurtles through space at thousands of kilometers per hour, an approximate reading will be enough for you to aim your camera at the general area. Expose several pictures with your best and tested exposure times and do so over a period of 3 or 4 hours. *Do not* change the target area. You should aim at the same point in the same star field (same R.A., same declination). Then start your blinking machine and put in, say, the first and third slides to start. Not only will the light of even a faint comet have started to build up on your film, making an otherwise insignificant and perhaps invisible object visible to your camera, but the jumping spot of light will give away its position in a flash. Having captured the new object, you can bring your longer lenses to bear on your target, or your telescope when you get one.

Don't forget your local television station or the press if you have a good picture. Video cameras have slide adaptors and your 35-millimeter shot just may be on the airwaves next evening on the 6 o'clock news, or in the Sunday paper on the following weekend.

The ultimate lure of the blinking discovery machine, however, is the very real chance it presents of making a discovery. No longer is this magic realm of opportunity denied to beginners and limited to the experts. Anyone can play, young or old, skilled observer or dabbling beginner, devoted stargazer or hit-and-run photographer. All that is required is the self-discipline which will assure photographic coverage of an area, however small, at certain definite intervals.

No telescope is needed, only the camera and the lenses which were already discussed. Use a well-aligned tracking mount which ideally should stay in one place or be located in such a way that the mechanism and the camera can be attached quicky to a permanent pier with a minimum of effort during the photo sessions and then removed for safe storage. Good steady work at the camera, combined with diligent observation on the projection screen, will make year-round observing into a rewarding pursuit. Instead of neck-craning stargazing searches in freezing winter weather, comfortable projection screen surveys of the skies can be made. Objects far beyond the reach of the unaided human eye can be reviewed in restful interior comfort from an armchair on cloudy evenings or on sunny Sunday afternoons.

You may not even be very familiar as yet with the area you photographed. However, familiarity with the star fields will rapidly increase as the photographs are repeatedly projected on your screen and the atlas charts compared with the pictures. To start, you need not even know the constellations precisely nor be aware exactly 'which star is which', or 'what belongs where'. In order to find that which does not belong at all, or to make a discovery, all you need to do is to photograph and then 'blink' your pictures.

Even advanced amateurs, however hard they may try to impress you with their knowledge of the skies, *cannot* be familiar with celestial territories at magnitudes fainter than 6 or 7. We already know that they would have to have memorized the positions of thousands of stars.

Narrow-angle long-exposure photography through long telelenses or telescopes will open vast new areas to advanced amateurs also. Such probings to fainter magnitudes at greater depths of unexplored space may also be rewarded with new findings such as supernovae in distant galaxies.

Let me share a secret with you. Early in 1976 when I had just started blinking, I compared a series of slides which I had taken in the region of Virgo. As usual I had used the inner cardboard tube on which toilet paper is rolled, held it to my eye to black out peripheral parts of the pictures as they flashed on the screen, when suddenly – would you believe it – there was motion there. I was riveted to my chair. Back and forth it went, unmistakenly an object in motion, a star in transit. My fascination was complete, my involvement total. I had made a discovery! Even the fact that I had bravely estimated the object at a sizeable magnitude 6 (comparing it to other objects in my charts) did not deter me from believing that somebody else might simply have missed this relatively large and bright newcomer. To me it was a heavenly body which must have escaped other observers.

Allow me here to boast my early friendship with Dr William Liller, then of Harvard University, co-author of this book. Our acquaintance stemmed from my accidental shooting of the series of sequential prediscovery shots of Nova Cygni 1975, on which he did the research. It was Professor Liller in Boston whom I called in the first flush of excitement after my discovery. 'I've found something moving with my blink comparator', I told him. Proud of the homework I had done, I added 'The object is about sixth magnitude and very near Lambda Virginis'. (I had to look up the Greek letter also to ascertain that it indeed was the letter lambda.) 'The object moved about 1½ degrees in the 9 days between photographs.' Bill was very considerate in the way he broke the news to me: 'I will now call you Sir William Herschel', he said, 'You have just discovered the planet Uranus'.

It may not have matched or exceeded the excitement of my Nova Cygni photo event, but the sense of exhilaration, even achievement, was total. I had discovered – independently – my very first own planet. No matter that it had been discovered 200 years earlier. If nothing else, it proved the effectiveness of my PROBLICOM (PROjection BLInk COMparator).

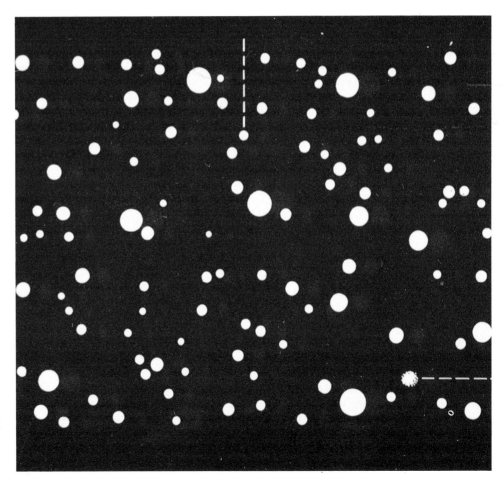

Fig. 16.1. *The Trifid Nebula, M20, in Sagittarius.*

CHAPTER 16

A night in a mountaintop observatory

In a book which emphasizes the joys and the scientific values of using the simplest of cameras to make astronomical observations, it may seem unfair to devote a chapter to the delights of using the world's largest telescopes. However, going to a major observatory and having use of its facilities for a single crystal-clear night is such a fantastic experience that we would like to share it with you. It might also give you some ideas about how to go observing, what to wear, how to select a good site, or building some semi-professional accessories of your own (see also Chapter 19). So, in your imagination, pack your warm observing clothes and flashlight and come along with us. Be patient just a little bit longer: in Chapter 17 we will suggest that you buy a telescope.

Site selection and telescope building

For obvious reasons, the best optical observatory sites are remote, high and dry. At an altitude of 2300 meters, one is already above one-quarter of the Earth's layer of air. Most of the water vapor, smoke particles and wind-blown dust lie below. At 5500 meters, half of the atmosphere is underneath you, and the thinness of the air makes breathing difficult. The highest observatory in the world at just over 5500 meters is located on a Bolivian mountain called Chacaltaya, but most of the research carried on there deals with cosmic rays, not light rays. The highest optical observatories in the world are on Mauna Kea, Hawaii (Fig. 16.2), and Pic du Midi, France, both at about 4000 meters. The majority of optical sites are located at altitudes between 1500 and 2500 meters.

Because weather obviously plays an important role in site selection, it is not surprising that three of the largest observatories in the world are now situated within 120 kilometers of each other in the foothills of the Andes mountains in the Atacama Desert of Northern Chile, one of the driest regions of the world. Other major observatories operate in the United States Southwest, and some of the earliest clear-weather sites were in California at Mount Hamilton (Lick Observatory), Mount Wilson and Palomar Mountain. Within Europe, Spain, Southern France, Northern Italy and the Canary Islands now have large telescopes. Still, because there is often a patriotic desire or a political need to have a national observatory located within a country's boundaries, we find important telescopes located in less-than-ideal climates such as in Great Britain, Japan and West Germany. Similarly, many universities recognize the importance of having moderately large telescopes nearby so that astronomy students can get the training they need without traveling long distances.

Who pays for the telescopes? In the United States many of the newer very large telescopes were built and are operated nowadays with funds from government agencies such as the National Science Foundation and NASA. For the use of American astronomers, the NSF supports the operation of four national facilities with large telescopes, two optical and two radio: the Kitt Peak National Observatory near Tucson, Arizona; the Cerro Tololo Inter-American Observatory near La Serena, Chile; the National Radio Astronomy Observatory in Greenbank, West Virginia; and the National Astronomy and Ionospheric Observatory near Arecibo, Puerto Rico (Fig. 16.3). KPNO and CTIO both have nearly identical 4-meter reflectors plus a battery of smaller telescopes starting at 84 inches (KPNO) and 60 inches (CTIO). NRAO has steerable parabolic radio reflectors of 300-foot and 140-foot diameters, and two large antenna arrays, one, very large, in New Mexico, used for detailed study of the brighter radio sources. At Arecibo the largest radio dish in the world scans the sky, a stationary 1000-foot monster that is basically a spherically shaped valley lined with wire mesh. A special receiver at the focus can be moved to follow the radio image of an object that moves across the meridian.

The world's largest optical telescope, the Zelenchukskaya (Fig. 16.4) with its 6-meter mirror weighing 38 tonnes, is located in the Caucasus Mountains of the Soviet Union; and at several other sites in the Soviet Union, identical 2.7-meter telescopes are available for local astronomers. Because of the generally poor weather for optical astronomy, the British and Dutch have in the past tended to emphasize radioastronomy in their respective countries, and the great steerable reflectors near Manchester, England (250-foot) and Dwingeloo, The Netherlands (25-meter) have produced many important findings. More recently, West German astronomers have built a 100-meter steerable dish in Bonn, and in Australia the Parkes 231-foot paraboloid has been the largest radiotelescope in the Southern Hemisphere for many years.

Recently, the British and the Australians have combined resources to build an Anglo-Australian 3.9-meter optical tele-

Fig. 16.2. (above) Vulcanologists are generally of the opinion that Mauna Kea, shown here, is in no danger of erupting again. At 13 800 feet, the summit makes an excellent site for an observatory. At the present time there are four telescopes larger than 2 meters, and more are planned.

Fig. 16.3. (facing page) The radiotelescope at Arecibo, Puerto Rico, was made from a natural bowl over 300 meters in diameter. This depression was accurately shaped into a sphere and then lined with a wire mesh. To follow celestial motions, the receiving antenna can be moved. It also corrects for the distortions produced by not having a parabolic surface (see Chapter 4). The Arecibo Observatory is part of the National Astronomy and Ionosphere Center which is operated by Cornell University under contract with the U.S. National Science Foundation.

scope situated at Siding Springs, Australia (Fig. 16.5), while the English, and the French and Canadians, have put into operation telescopes of similar size on Mauna Kea, Hawaii. Probably the best organized overall effort is the European Southern Observatory (Fig. 16.6), operated by a consortium of 11 nations and located 100 kilometers north of the observatory of Cerro Tololo in Chile. ESO's telescopic inventory includes a 3.6-meter reflector, a score or more smaller instruments beginning at 2.2-meters and a 1-meter Schmidt camera. This latter telescope plus a similar one in Australia has recently mapped the skies of the Southern Hemisphere in several colors as the Palomar Schmidt did in the Northern Hemisphere in the 1950s and is doing again.

The title of the 'most famous telescope of all' is still held by the 200-inch reflector on Palomar Mountain. The vision of George Ellery Hale, the great American astronomer of the early twentieth century, this giant instrument, completed in 1947, is the product of private funds mainly from the Carnegie fortune. A few large telescopes continue to be financed privately such as the du Pont 102-inch located on Las Campanas in Chile, and the great era of building telescopes with private money that began at the end of the nineteenth century continues. The names of the benefactors are remembered in the names of the observatories: Yerkes, Lick, Carnegie and others. More names will be added in the future.

A recently built large reflector, the multi-mirror telescope (MMT) on Mount Hopkins south of Tucson, Arizona, was financed jointly by the Smithsonian Institution and the state of Arizona. The novel design of this telescope, with six 72-inch mirrors operating as one, may very well become the design of the future since mirrors larger than 6 meters in diameter will be extremely difficult to cast, shape, transport and support properly in a telescope tube. The light-gathering power (see Fig. 2.6) of the MMT, equivalent to that of a single 178-inch mirror, is impressive; this telescope now serves as a model for new, giant telescopes already in the design stage.

Fig. 16.4. (left) The great 6-meter telescope of the Soviet Union, the world's largest. Its official name, Bol'shoi Teleskop Azimutal'nyi, translates literally to Big Alt-Azimuth Telescope. It is located on 2100-meter Mount Pastukhov near the city of Zelenchukskaya in the southern Soviet Union.

Fig. 16.5. (facing page) The 3.9-meter telescope known as the Anglo-Australian Telescope, is located on 1200-meter Siding Spring Mountain in Australia. It went into operation in 1975.

The 'typical' observatory

While the equipment of most major observatories seems to be dominated by a single large telescope, the more correct concept is that there exists a broad selection of instruments ready to serve the astronomer in many ways. It is not necessary to use the largest available telescope for many observing programs, and in fact to apply successfully for its use, one has to show that the largest aperture instrument is the only one capable of carrying out the intended research.

How many of what kinds of telescopes are available to the astronomer at a modern observatory? Consider, for example, the Cerro Tololo Inter-American Observatory in Chile: what telescopes are there and how are they used? The largest is the 4-meter reflector, followed by telescopes (all reflectors) with mirror diameters of 1.5 and 1 meters, and 36, 24 and 16 inches plus a Schmidt camera with a 36-inch mirror and a 24-inch diameter correcting plate. Three of these instruments are owned by other institutions (the Yale University 1-meter, the University of Michigan Schmidt, and the Lowell Observatory 24-inch), and they make a substantial portion of their observing time available to guest investigators from other institutions.

With the exception of the two largest instruments and the Schmidt, the CTIO telescopes are operated at the Cassegrain focus (see Chapter 4) where one usually finds a spectrograph or a photometer of some kind, although all can be used for direct photography. (A photometer is, in effect, a supersensitive light meter; see Chapter 20).

For most telescopes, there are two or three secondary mirrors available, providing different effective focal ratios. The usual ones are f/7.5 and f/13.5, although the Lowell Observatory 24-inch, occasionally used for planetary photography, also uses f/40 to provide a very long focal length and a correspondingly enlarged scale at the focus. The Michigan Schmidt, which is used exclusively for photography, operates at f/3.5, thereby making it possible to take wide-angle sky photographs at high speed.

When the Moon nears its full phase and no longer permits astronomers to study faint stars and nebulae, telescopes are frequently used with high-dispersion spectrographs to analyze in great detail the light from the brighter objects in the sky. Just about all large telescopes can be operated at a special Cassegrain focus known as the coudé focus, a large focal ratio arrangement

Fig. 16.6. The European Southern Observatory in northern Chile. More than a dozen telescopes make up the inventory of this well-equipped observatory. Almost exactly 100 kilometers to the south is the Cerro Tololo Interamerican Observatory.

providing a stationary observing platform on which large and complex instruments, usually spectrographs, can be mounted. Sometimes the cost of the coudé spectrograph alone exceeds that of the rest of the telescope.

It is during the dark phase of the Moon when the observatory is at its busiest. Only then can the astronomer work on very faint objects buried in the always present background glow of the sky. This dim illumination emanates from myriads of even fainter objects – stars, nebulae, galaxies – and the weak light produced high in the Earth's atmosphere by cosmic rays. Also contributing is the always present but usually weak Aurora Borealis (or Aurora Australis in the Southern Hemisphere).

One of the most exciting places for the astronomer to be is in the prime focus cage of one of the giant reflecting telescopes, where he literally rides the tube through the sky, peering down on an immense piece of glass held rigidly but delicately at the far end of a rugged framework of steel (Fig. 16.7). The entire 30- or 40-tonne mass with its main horseshoe-shaped bearing literally floating on a thin film of oil is moved by a fractional horsepower motor with a precision far greater than an expensive watch. Expertly and effortlessly, a night assistant manipulates the controls of the giant instrument, directing it at the chosen place in the sky; then the encapsulated astronomer takes over and begins making his measurements or taking his exposures, guided all the while by some distant star and adjusting the telescope's position slightly from time to time. Rarely are the corrections, made necessary by minute flexures in the tube or by small, unpredictable fluctuations in atmospheric refraction, larger than a few tenths of seconds of arc.

With most of the newest large reflectors much of the romance of this kind of observing has been replaced with television acquisition, computer-controlled data-taking and automatic guiders. Still, many astronomers find photography and the presence of the observer essential for their research.

So the work continues through the night. In wintertime no heat can be turned on in the dome because the rising warm air would spoil the seeing. Consequently the astronomers and night assistants must engulf themselves in layers of warm clothing. With few interruptions they work non-stop, giving up only when dawn – or clouds – begin to interfere with their measurements.

The sounds of an observatory night are few, distinctive and fascinating: the hum of a transformer buried in the electronic equipment, a computer printing out data, the occasional click of a relay as the observer presses guide motion buttons and, once in a while, the sudden rumbling of the dome as it is rotated to catch up with the moving sky. Often this observatory music will be mixed with more conventional music provided by a tape recorder brought along to make a long night pass more pleasantly. Or sometimes you can hear the crackle of a short-wave radio bringing in an all-night station from some distant, foreign city. Usually not much is said by either the astronomer or night assistant, since almost continuous attention must be given to the guiding and data-taking throughout the precious night.

With the coming of dawn, the astronomers close up the domes, turn off their equipment, and head for the dormitory where they will sleep until early afternoon when they rise to eat and look over the last night's data. At nightfall, it all starts again, weather permitting.

In principle, anyone is eligible to use any of the telescopes; the only prerequisites are experience and a worthy observing program. Once or twice a year a scheduling committee of experts evaluates dozens of observing requests and selects what they believe to be the best programs. Specific nights are then assigned to each chosen astronomer, notifications sent out and acceptances returned. In the weeks before the appointed time, the preparations are made: a list of objects to observe must be drawn up, coordinates 'precessed', finding charts prepared, and exposure times determined. Every minute of the valuable observing time is carefully planned. Astronomers far outnumber the available telescopes and observing time is a precious commodity.

Getting to a distant mountaintop observatory can be expensive, and to make it possible for all scientists throughout the United States to use the national facilities, the National Science Foundation pays most of the travel expenses. It also supports a staff of astronomers, computer programmers and equipment specialists at each of the National Observatories; they divide their time between their own research and seeing that the telescopes are properly maintained and utilized. A sizeable group of expert technicians is also employed because keeping a large array of complex precision instruments in first-class working condition 365 nights of the year requires almost daily maintenance and testing. In addition there must be on hand a full stock of supplies – photographic plates and processing chemicals, cryogenics to cool photocells, magnetic tape and paper for the computers, and seemingly thousands of other items which the guest investigator cannot always be expected to carry with him but may possibly need during his stay.

Always there is the weather to contend with. Even in dry Northern Chile there can be clouds, although during the summer months – December to March – they are rare. Still, on occasion an astronomer will travel all the way to a 'clear-weather site' only to spend most of his nights reading and writing, talking to other

astronomers, playing cards and preparing a future observing program. Usually, however, the astronomer returns home after 10 or 15 nights of observing with enough data to keep busy until the next time which may be 6 months later, or 6 years, depending on the nature of his observations.

Pity, if you will, the poor computer-oriented observer who spends the night sitting in a warm, well-lit, comfortably heated, luxuriously furnished observing room next to the main dome. Gone is the tingle of the glory of the sights of the night sky. But pity also the astronomer who comes to his unheated dome unprepared for the rigors of a winter night. Rare is the astronomer living or working in so-called temperate climates who has not at some time or other felt bone-chilling cold, sitting ill-prepared beside a telescope warmed only by the glow of the stars. At those times when the gentlest breeze can feel like a raging Arctic blizzard, one quickly appreciates immensely the slight protection given by a dome or other telescope housing, no matter how rudimentary.

The astronomer soon learns the fundamental rules to prevent the aching agony of cold from completely destroying the pleasures of outdoor observing (Fig. 16.8). Wear many layers of loose-fitting clothes; don't begin the night with clammy feet moist from the day's labors; don't allow fingers to make direct contact with metal (use gloves or plastic knobs and handles); and above all keep the upper body warm. Cold hands and feet are bearable but once the chest is frigid, you're done as far as observing is concerned.

Electrically heated jackets, trousers, gloves, socks and mittens exist but are hard to find. Try military surplus stores or specialty shops, or simply wrap an electric blanket around you. Try sitting on a heating pad. Don't use too much heat though or you'll spoil the seeing.

Fig. 16.8. An astronomer about to begin a long winter's night's observing ought to be well prepared. Drawing by Lucille (Mrs Ben) Mayer.

Fig. 16.7. For telescopes larger than about 3 meters, the observer or his equipment can ride on a prime focus capsule located directly at the main focus of the parabolic primary mirror. Here we see the upper section of the 4-meter reflector (and this author) at the Cerro Tololo Interamerican Observatory.

Fig. 17.1. The Dumbbell Nebula, M27, in Vulpecula.

CHAPTER 17

Go ahead, get that 'scope . . .

If you have – unlike me – controlled yourself and have not bought a telescope yet, then many wonderful opportunities and choices present themselves to you now. They should all include an equatorial drive, a 'stop the world' mechanism. This part of the system you may already own. It constitutes the critical half of any telescope.

With whatever telescope system you chose – a variety are discussed below – buy an inexpensive 'camera attachment clamp' right away. This will permit you to mount your camera on the barrel of your telescope so you can begin (or continue) your basic starlight collecting. This will make use of the valuable equatorial drive of your new acquisition. Such photography with your camera riding 'piggyback' on your motordriven telescope will give you predictably good photographs as you acquaint yourself with the appointments of your new system. Walk before you run. Spend extra time learning polar alignment which need not be too critical for such 'piggyback' photography but must be accurate for photography *through* the telescope. Follow the general instructions for polar alignment which will be included with your new telescope. These will suffice for basic picture-taking.

TO AVOID DISAPPOINTMENTS PLEASE TAKE YOUR FIRST PHOTOGRAPHS IN THE PIGGYBACK MODE, i.e. WITH YOUR CAMERA RIDING ON TOP OF THE DRIVEN TELESCOPE.

Otherwise your purchase may end up – like so many – in the 'For sale' classified section before you even had a chance to capture some of the magic of the night sky.

A very basic decision should be made early: do you want to refine your photography or does visual observing intrigue you? Don't forget professionals rarely look through their instruments because film can collect starlight best. While it is possible to attach cameras to just about any telescope, there are the special Schmidt optical systems which are designed only for photography. It is impossible to 'look through' a Schmidt photographic telescope but for this drawback there are ample compensations: where all viewing telescopes, reflectors or refractors, have speeds anywhere from f/4 all the way up past f/15, a Schmidt 5-inch or 8-inch captures light at a stunningly fast f/1.6 or f/1.5. This makes such instruments truly effective, optically perfect recording cameras. They can photograph in 2 or 4 minutes that which regular cameras attached to any other kind of telescope would take half an hour to shoot. We are talking photography so there is a trade-off again. Schmidts are indeed telescopic in nature but the focal length of the ones on the market already mentioned are 230 and 305 millimeters, respectively, nowhere near as great as regular viewing telescopes. This automatically means that while the field which one photographs is greater (and every object needle sharp), most individual clusters, galaxies or spirals *cannot* be made to fill the picture. This is where enlargers come in with which one can pick out areas from the negative to blow up and dramatize.

When comparing a 5-inch Schmidt (where the 5-inch refers to the diameter of the front 'correcting' lens) with a regular telelens of 300-millimeters focal length, you may well ask whether the standard long lens which is compatible with your camera is not the better buy. The answer must be yes and no. *Yes*, the regular 300-millimeter is the more practical buy because it can be used in daytime for terrestrial photography where the Schmidt cannot be used for anything but night starshooting. *Yes*, the standard telelens is the superior value if you think in terms of easy attachment to your camera. Films run through the system in the conventional manner, are advanced by pushing the lever a frame at a time which allows rewinding the exposed film into a cassette before processing. The Schmidt requires slightly more difficult film handling. *No*, a regular telelens is not more desirable than a Schmidt because the quality of the star images cannot compare to those on Schmidt films. Enlargements of fair quality are possible from standard telelens negatives or transparencies but are superior from the Schmidt. Finally, *No*, a conventional bayonet or screw-mount camera lens, however long, simply *cannot* be obtained having an f/speed even approaching that of a Schmidt. The reason, as already outlined in Chapter 3, is very simple: as a funnel standing in the rain collects raindrops from a great area and concentrates them into a small opening beneath, so the Schmidt gathers light through the entire sizeable diameter of its barrel opening and, with a concave mirror at its bottom, focuses all the photons into a small area where the film itself is located. This is a direct and fast arrangement which – by the way – has only been commercially available to amateurs since about 1970.

We must not allow ourselves to forget the rather basic truth which visual observers cannot debate, that only through photography and the patient collecting of light on film do the major sky attractions become visible. Therefore, the preference for equipment which will facilitate such photography becomes an important alternative.

If astrophotography remains your preferred method of observing, if collecting photons is going to be your principal pursuit, then consider either of the following:

Telelens approach: Buy a 135-millimeter, fastest telelens or a 300-millimeter. Don't go to anything longer because the speeds fall off rather drastically and the sheer weight of such lenses makes them cumbersome and difficult to handle. You may also wish to consider the 'reflector' lenses.

Schmidt approach: Buy a 5-inch diameter Schmidt or, if you can swing it, one with an 8-inch diameter. Their speeds, as already stated above, are superfast, and learning to use these instruments is relatively easy. At this time only a few telescope manufacturers produce such systems but other makers are bound to follow (see Appendix).

Before reviewing the second alternative, that of observational astronomy, it should be stated once again that the only reason more amateurs do not practice photography is because they mistakenly think that it is a difficult pursuit. This is not true.

As concerns telescopes for visual observation, the available selection of instruments is virtually endless. At the outset the exciting and rewarding possibility of 'building your own telescope' must be stressed. All over the world there are amateurs whose common bond is mirror-grinding and telescope-making. There are clubs devoted to the furtherance of this art and meetings at the local, regional and national levels where telescope-makers show their inventive handiwork and compare their construction methods and optical results. Many books have been written on the subject and some are listed in our Bibliography. In Chapter 18 we will describe briefly the art and science of mirror-making.

Still, here too you will want more than book knowledge. Go join a group dedicated to telescope-making. See how even a 12-year-old who has made his own primary mirror will be proud to show you, and share with you, the excitement of 'doing it yourself'. Quite apart from the tremendous savings in money which are possible through such an approach, you will gain access to real technical knowledge and optical know-how. Invariably persons with extensive practical experience form the core of telescope-making groups and everyone without exception can be counted upon to be generous with advice, even help. If you're lazy as I am you will want to shop for a ready-to-gaze telescope. No one will blame you. There are often more 'commercial' telescopes at star parties than home-built ones. Frequently there will be modifications to such factory-built scopes and you too will add features of your own to whatever make of instrument you purchase.

As with cars you will develop loyalties to certain manufacturers' products. The ease of compatibility of mounts, drives and telescopes all from one source makes for practicality. Still, as with stereo equipment, it is possible to purchase components, such as one kind of mount and a telescope of another make which can be combined. Eyepiece lens barrel diameters are standardized at $1\frac{1}{4}$ inches in diameter and oculars and attachments come in a wide range of matching types.

Generally speaking, Newtonian telescopes (see Chapters 4 and 18) are an excellent value when purchased commercially. They can give you maximum aperture for minimum cash outlay. These of course are reflectors and speeds of f/4 are possible with mirrors 6 inches (24 millimeters) in diameter and larger. Such telescopes are not overly heavy even if of large diameter and may well be the most popular among amateurs. Cassegrain telescopes and so-called catadioptric systems come next. These have primary mirrors at one end which fold the light beam back to a second mirror to return it through a hole in the prime mirror where the image is presented to the eye or camera. Next come the refractors which date back to the ancient spyglasses of mariners. They invariably form part of any of the foregoing types of telescopes for which they serve as viewfinders. Large refractors employ only lenses and are the oldest types of telescopes. Most standard camera telelenses are in fact refractors with a film where the eye would be if the instrument were a telescope. On the one hand, good refractors with larger diameters can be excellent visual aids; on the other, these long-barreled descendents of Galileo's looking glass, as already stated, are often temptingly merchandized by promotion-minded stores as having capabilities beyond belief or credibility – or need. Generally speaking, in terms of dollars per inch of aperture, *good* refractors are far more expensive than any of the mirror types.

An important point must be made here relating back to astrophotography and the 'guiding' of long-exposure photographs, particularly where long focal lengths and high magnification are

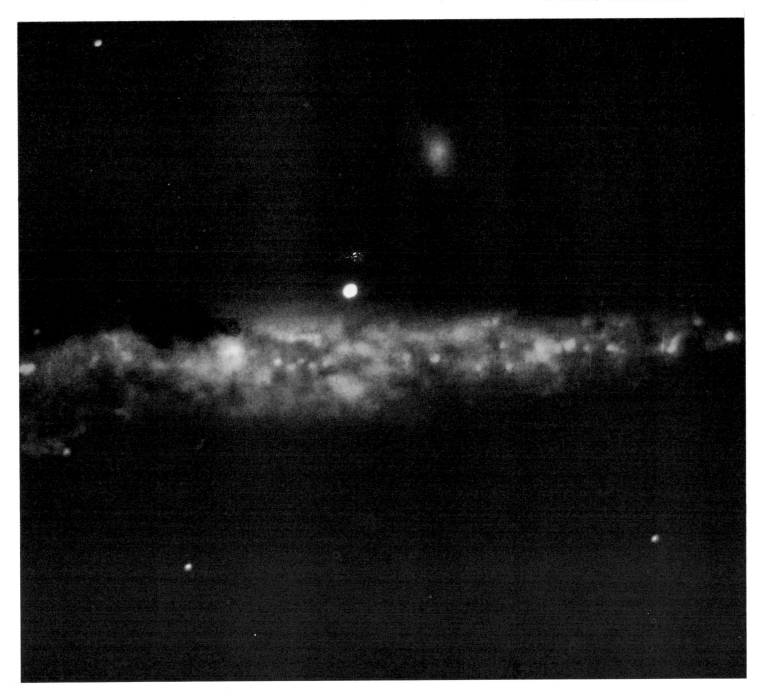

Fig. 17.2. NGC 4631, spiral galaxy seen edge-on.

involved. The truth of the matter is that when one aims for the high-quality photograph of which every amateur is capable, one needs a guiding system which can be attached to, or is part of, the telescope through which the pictures are taken. Guiding, as a rule, should be done with a good-quality optical system having a magnifying power greater than the photographing telescope itself. For both the 'telelens' and the 'Schmidt' approaches mentioned earlier, an inexpensive refractor can readily fill this need when it is attached to the photosystem. Other guiding systems use 'off-axis' methods, where stars on the edge of the photographic field are used to guide long exposures.

To explain the difference between motor-driven exposures of larger areas, as defined in Chapter 7, and 'guided' telescope exposures of small fields, a comparison can be drawn between a shotgun and a rifle. When skeet shooters aim for clay pigeons they use large-bore shotguns, follow the general direction of the moving target, and blast a great number of pellets into the air thus increasing their chances of hitting the fast-flying target. Marksmen, on the other hand, must rely on single shots with a carefully aligned gunsight. They slowly move high-powered weapons to keep a distant target carefully centered in the crosshairs of a gunscope.

A wide-angle shot, where a larger area is photographed, does not demand the accuracy of aiming which is required when we attempt to zero in on a single object which is tiny and very far away. For that we need precisely what the marksman must have – a telescopic sight with crosshairs.

By the mere addition of an eyepiece with crosshairs, any telescope can become such a sighting device. Since it would be virtually impossible to see the thin cross of hairs at night, we employ what is known as an 'illuminated reticle', a fancy name for two lines scratched at right angles into a glass which is placed in the line of sight usually in the eyepiece. A tiny battery-driven 'wheatgrain' lamp or LED (light-emitting diode), off to one side, 'edgelights' these lines and makes them glow very slightly in the otherwise black star field.

In the case of astrophotography, it is not always possible to center in the crosshairs the object which one is shooting. It may be too faint of itself to be easily visible. More yet, the object may in fact be invisible even to the aided eye and only the photographic exposure will produce it, as if by magic. For this reason it is quite customary, after aiming the photographing telescope at the point in the sky where the desired object is known to be, to target the sighting telescope on a good, bright 'guide star' near the selected object. If one holds this bright and convenient star centered in the crosshairs for the duration of the exposure, then the object which one wants to photograph is recorded without movement on the film in the focal plane of the main 'scope. To enable the photographer to sight on one star and shoot another object, small linkages are provided between the guide telescope and the camera which allows such separate sighting-in by merely turning a knurled ring or two.

Since most celestial objects do not move appreciably during the period of even a long exposure, the guiding is done for quite different reasons from those associated with moving targets. All motors which gradually move telescopes to compensate for the rotation of the Earth employ gear arrangements of greater or lesser sophistication and accuracy. The more expensive the drive, the more accurate and steady the motion usually is. Nevertheless, a certain amount of slippage or error can and does occasionally occur, which can be easily compensated for by electronically speeding up or slowing down the motor. Electric motors themselves are not the culprits of driving inaccuracies. Rather gears, worms, pinions and bending in the mounting can cause small errors which add up when seen on photographs exposed over long periods of time.

East–west corrections in Right Ascension are easily made manually through linkages or by pushing an advance or retard button on hand-held electrical devices. These are wired, through a drive corrector, to the drive motor. Corrections are made while actually looking through the illuminated crosshair eyepiece at the star upon which one guides and by holding that star neatly centered on the vertical hair.

North–south corrections in the declination axis are usually made mechanically by sensitively turning an adjusting screw. Declination corrections are almost always related to the degree of accuracy with which one has aligned one's telescope with the Celestial North Pole. If the system is perfectly or nearly perfectly sighted in, almost no corrections in this axis are needed.

Write off to the manufacturers whose advertisements appear in the various journals, compare features or better yet compare actual instruments and systems at star parties. Check and compare for yourself the optical crispness of a certain object in two different telescopes of nearly equal cost or diameter. Look for design and structural integrity. Avoid flimsiness of construction or slick, fancy features. Chrome features may look great on a boat but do little for a telescope. Above all, wherever possible, always compare apples and apples before comparing apples and oranges. Thus, check one Newtonian against another rather than weighing Newtonians against catadioptric systems. For telescopes, the name of the game is aperture. Equally important is a sturdy base to support the weight of an instrument and future accessories.

In your final choice, you will be completely on your own. Asking an owner or manufacturer of a certain type of telescope whether another might be superior is like asking parents whether there are any children around prettier or brighter than their own. You know the answer to that one. In time, when you too have become the proud owner of a system, you may find yourself carrying a picture in your wallet of your prized optical possession, ready to show to one and all at the drop of a hat! You will have joined the ranks of the happy stargazers.

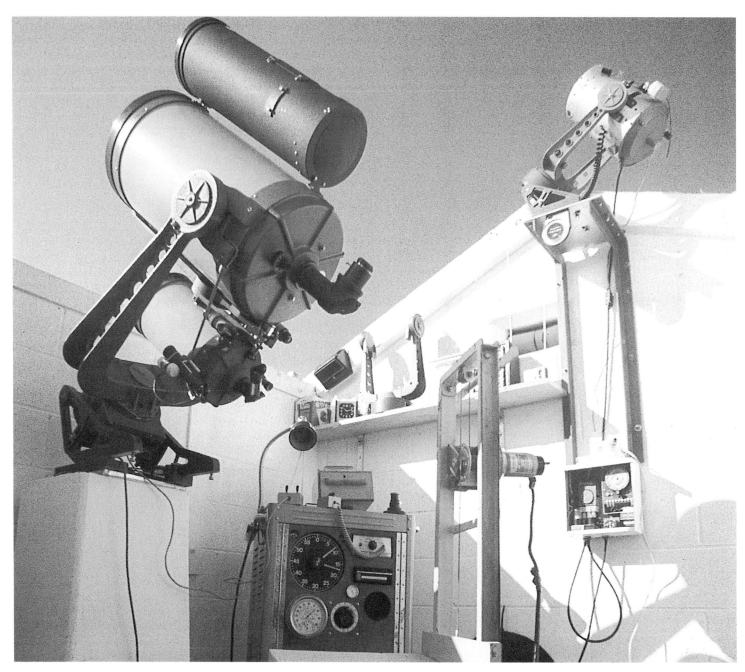

Fig. 18.1. *The well-equipped amateur can do astronomy at a level of quality that rivals professional astronomy. Here is Ben Mayer's observatory.*

CHAPTER 18

Choosing and using a telescope

Having just urged you to buy a telescope, we will now make some suggestions about what to look for, and how to use and care for your new acquisition.

Factors to consider when buying

As the previous chapter stated clearly, aperture is very much the name of the game. Next to the basic overall quality of the instrument, all else ranks lower in importance. These statements immediately lead one to ask, 'Why consider anything but a Newtonian reflector?', which as Chapter 17 pointed out, is by a wide margin the best aperture bargain. The answer depends upon the pocketbook: if you can afford to buy one, a catadioptric system or a refractor will require less maintenance and will take more abusive treatment than a Newtonian, and both will give you more focal length (which you may or may not want).

If you do end up purchasing a telescope with one or more mirrors, we again repeat this precaution: astronomical mirror surfaces are extremely fragile to the touch since, unlike a bathroom mirror, the reflecting coat of aluminum is on the front surface, exposed to the elements. An 'overcoat' of harder transparent material provides some protection for the aluminum, but not much. Don't touch it – or any optical surface.

For a telescope of a given size, the price depends primarily on the number of optical surfaces to be ground and polished, and secondarily on whether or not the light passes through glass which then must be pure and clear. After that, what you pay for is rigidity of mount (important), sophistication of overall design (less important), and attractiveness of finish (unimportant).

As far as the quality of the mirror is concerned, there is a straightforward way to tell what you are getting, and that is by what fraction of a wavelength of light the surface of the primary deviates from a perfect shape. Thus, if the 12-year-old who has ground his own tells you truthfully that his mirror is 'good to a quarter-wave', you should be impressed by his enterprise. The quality is passable, but just barely, and a telescope manufacturer should do better. An eighth- or tenth-wave is what he should deliver, and a twentieth is overdoing it unless the maker is talking about the secondary mirror of a catadioptric or Cassegrain where this kind of precision is necessary.

Coated optics increase the light transmission of a lens and, as mentioned, coating reduces the fragility of the reflecting surfaces of mirrors. Except for sealed systems, as in catadioptric

Fig. 18.2. A surprisingly powerful and enjoyable telescope can be home-made using only the simplest of materials. Dragan Mikesic, shown here, lives in Belgrade, Yugoslavia, and built this little gem of an instrument.

instruments, mirror coatings are worth the extra dollars.

If the effective focal length of the telescope is more than 0.5 or 1 meters, the mounting should be an equatorial-type and must be solid and rugged, not light and spindly. Those who wish to devote all their time to comet 'sweeping' may prefer to use an alt-azimuth mounting, likewise solid and rugged, like a 'Dobsonian' mount. Motor drives are a requirement for any instrument used for serious astrophotography, and are a great convenience for an equatorially mounted visual telescope. A good, serviceable equatorial mounting can be made of 2-inch or 3-inch cast iron pipe, especially if you live within 10 degrees or so of latitude 45 degrees (since 45-degree pipe angles are common) or, of course, zero or 90 degrees. A few degrees can be corrected for by tilting the main vertical member. Installing a drive is more involved, and buying a first-class factory-built motor-driven mounting might be the best idea if you are hoping to get first-class long-exposure photographs. Worm-gear drives, where a large, many-toothed gear is driven by a precision screw, will be more accurate than a driving system made of a chain of standard spur gears. Remember, too, that the smoothness of shaft rotation in a low-revolutions-per-minute synchronous (timing) motor is important, and that slow speed is usually achieved with a gear train built into the motor housing. A good quality motor is just as important as the quality of the mounting itself.

More and more often, the drives of professional telescopes are being converted to stepping motor drives. Stepping motors are precise since shaft rotation occurs in steps which are always exactly the same, but each step is achieved by sending an electronic pulse to the motor and the frequency of pulse rates can be accurately and easily timed and controlled.

Being able to vary the rate of your telescope drive is a blessing since the Sun, Moon, each of the planets, and the stars travel at slightly different rates through the sky. Commercially available systems provide separate settings for these objects. A pair of pushbuttons, one to speed up the drive, one to slow it down, makes possible precise guiding in Right Ascension. Declination drive motors also exist and make life easier, too.

Grinding your own mirror

As my co-author pointed out in Chapter 17, many people have ground their own mirrors with spectacular success, and we encourage you to do so if you have even the slightest knack with tools. Buy a good book on the subject if you are interested. Meanwhile, here are some general comments.

The wavelength of yellow light is about 590 nanometers (24 millionths of an inch). A good machinist can turn out a metal piece on his lathe or milling machine good to about ten of these waves, but an optician finishing a fine mirror surface, accurate to a tenth of a wave, does a hundred times better and worries about one or two millionths of an inch. Grinding and polishing an 8-inch mirror to this precision takes a reasonably well-coordinated but untrained person about 40 hours, and the cost of a do-it-yourself kit for an 8-inch is about the same as that of an inexpensive camera. Low cost may be an attractive incentive for making your own mirror.

The principle involved is simple. One starts with two

Fig. 18.3. Grinding a telescope mirror requires patience, a modicum of talent, and not much money. Denni Frerichs of the Chabot Science Center, Oakland, California, is shown grinding her own.

identical discs of glass, usually pyrex or, in some cases, more exotic materials like Cervit, UBK-7, or fused quartz. One slab, the tool, is firmly attached to a solid, waist-high pedestal. After spreading a teaspooonful or so of wet grinding compound, usually carborundum, on the tool, the hand-held piece of glass is rubbed back and forth across the stationary piece. During this process, the person doing the grinding slowly moves around the pedestal while gradually rotating the upper slab in the counter direction. The natural result of this procedure is that the surface of the fixed glass tool develops a convex shape while the other disc turns concave and spherical. Finer and finer grinding compounds are used as the intended degree of concavity is reached and a smoother surface is desired. Then follows the parabolizing process which requires deepening slightly the central region of the mirror and using a more localized rub with a wax or pitch surface added to the tool. Frequent optical testing employing nothing more than a razor blade and a pinhole light source allows the telescope-maker to keep a precise check on the progress. When the mirror has finally reached the intended focal length and accuracy of shape, the surface is polished with optician's rouge and then sent away for aluminizing (or it can be silvered at home).

Amateur telescope-makers are a justifiably proud lot and have formed clubs in some large cities, usually with the cooperation and encouragement of the local science museum or planetarium. To have created an optical component with a surface accurate to a few millionths of an inch is an achievement worth bragging about, to say nothing of the fine views of the astronomical world that one can get.

The care of a telescope

Aside from the special care that the optical surfaces require, which are summarized below, there are no precautions that one need take other than those of common sense. Dry and dust-free conditions will help keep metal surfaces from corroding and moving parts from grinding together. There will be times when the humidity rises to the point where condensation occurs and the entire telescope becomes dripping wet, finding charts and record books turn soggy, and in general observing loses the appeal of warm and drier nights. However, at these moments, the seeing is apt to turn spectacular and to close up would be unthinkable. Afterwards, application of a soft, dry cloth to the dripping non-optical parts is essential. Any fogged optics *must* be allowed to dry by evaporation, helped if necessary by a gentle application of warm air from a hairdrier.

Your telescope should also receive special attention after those nights when a cold front has turned the sky sparkling clear but has brought with it gusty winds which kick up dust and grit. Gear teeth and axis bearings should be brushed clean with a soft, dry brush.

Now some words about the surfaces of mirrors or lenses. Again the cardinal rule to follow is: **Never touch!** Blow on them if you must, preferably with one of those cans of dry air or inert gas, but never, never allow fingers or noses even to come close. Apply a soft lens brush, lens cloth or lens tissue only when you feel you absolutely have to. A correcting lens or aluminized mirror which looks scandalously filthy actually continues to transmit or reflect a surprisingly large percentage of light. In fact in the days before antireflection coatings, it was found that a little bit of dust and grime on a lens actually improved its efficiency.

When the time finally comes when there is no other recourse but to disassemble and clean the optics thoroughly, use liquid soap or detergent, preferably unscented; great quantities of pure water, distilled if possible; and big wads of good quality cotton. Let parts dry by evaporation in a warm, dust-free environment. *Remember*: every scratch, even those barely visible to the naked eye, causes scattered light and washes out the image. Furthermore, one is continually faced with an irreversible situation: scratches may come but they never go away – not unless the mirror or lens is refinished.

Eyepieces

As a general rule, the more an eyepiece costs, the better it is. One exception is a 'zoom' eyepiece, which cannot be enthusiastically recommended unless you have narrow field-of-view applications and plan only to look at objects which are at least moderately bright. We should also add that a 15-cent magnifier can serve at a pinch, but does leave much to be desired.

An eyepiece turret which keeps to hand three or four 'oculars' of different focal length provides convenience but also additional maintenance to keep the hub from getting too loose or too tight on its rotating axis. A padded box with hinged lid and places for each of your eyepieces is recommended if you are happy using nothing more than a sliding tube or rack-and-pinion assembly for the viewing end of your telescopes. Perhaps the best of all storage places, though least elegant, are pockets – dry, warm, clean pockets. Your body warmth reduces the possibility of lenses fogging over, especially if the night is humid or the face slightly damp from perspiration.

If you are in the market for eyepieces, review the section in Chapter 4 on the subject before deciding on focal lengths. Your own particular telescope may set some restrictions that you should know about. Otherwise aim at getting a set of four oculars ranging from 8 millimeters to 50 millimeters. An additional 25-millimeter eyepiece with illuminated reticule (or cross-hair) will be useful, if not essential. Like shoes or neckties, your collections will grow with time; you will have favorites and then there will be the others seldom used.

Photography with your telescope

Let this chapter reinforce the importance of starting with piggyback photography. Clamp your 35-millimeter camera rigidly onto your telescope. The focal length of the camera lens is not critical. Start with a lens of 50 millimeters focal length and then, if you have one, a telelens. As noted earlier, shooting with a telephoto lens almost requires a guide telescope. Your new telescope, equipped with an eyepiece with illuminated reticule and a motor-driven equatorial mount, is the perfect tracking system. Now you can take beautiful long exposures while guiding all the time on a convenient bright star. There are dozens of inviting subjects: Milky Way star fields, the nearer galaxies, rich star clusters, the larger gaseous nebulae, and comets. Then, of course, there are the thousands of variable stars whose magnitudes can be easily determined from such photographs since around each variable you will have numerous comparison stars. All these are within easy reach of your camera riding piggyback on your telescope.

Later, after you have become experienced at polar alignment and guiding, you will be ready to embark on the fascinating program of photography through the telescope. The Moon should be your first subject; it's bright and big. Later, go for the more challenging celestial objects. Planets are bright, but to capture the fine detail of Saturn's rings or the belts of Jupiter requires good seeing and a fine touch at the telescope controls. Larger objects like open clusters, globular clusters, gaseous nebulae, even galaxies, are fainter and require long exposures to bring up the delicate wisps and star chains, but now you are ready to make the necessary long exposures to capture these astronomical gems.

Besides accurate guiding, one needs a carefully focused system to get the best photographs. In Chapter 19 you will find some excellent tips on the best way to focus. Follow these steps carefully. Also, refer to Chapter 17 for a clear description of the best guiding arrangement to use; several successful systems are now marketed (and advertised) and you should consider each carefully to know which is best for your application.

Telescope accessories

The serious astrophotographer will find that the weight of the tail-end accessories grows quickly as his appetite for additional and more elaborate equipment grows. One may, for instance, feel the urge to experiment with photometers or astro-TV (see Chapter 20), or with spectroscopy (see below), or merely add on 'a few extras' like filter-holders, illuminated reticules, tele-extenders or telecompressors. In either case, good telescope stability is required. Both the basic equatorial mounting and the accessory holder should be rugged; when making the original purchase or design of a telescope, think in terms of hanging on 4 or 5 kilograms of additional equipment at the focus. Cold cameras, for example, can weigh that much. A convenient tail-end arrangement that is found on most professional telescopes is a focusable platform, rather than the standard rack-and-pinion mounting. The accessories can be firmly bolted to this platform, providing a much better arrangement than the simple sliding tubing. One type of mounting plate that can be purchased is a laboratory 'scissor jack'; an inexpensive substitute can be built using a pair of 10-millimeter aluminum plates held in place by three or four hefty screws and heavy-duty springs.

If you want to increase the effective focal length of your telescope, buy a tele-extender, sometimes called a Barlow lens. It is nothing more than a *negative* achromatic lens which is placed in front of the telescope focus and effectively increases the focal length of the objective by delaying the convergence of the incoming light rays.

You may, on the other hand, want to decrease the effective focal length of your telescope. You will find this to be an excellent idea for photography of the more extended objects including many clusters, nebulae and comet tails. Then you will use your camera with a 'telecompressor' lens, which is simply a *positive* achromatic lens placed in front of the main telescope focus. Using a telecompressor effectively speeds up your system.

Let us consider the automatic guiders. One can buy (or build) a comparatively simple guiding device, which comes with a photocell plus associated electronics and can be slipped into a standard eyepiece-holder. The idea behind one of these marvels is to block off half the light of a star image with a knife-edge at the focus, and have an electronic warning system which, in effect, pushes the eastwards or westwards button of your telescope whenever the light intensity of this half-star increases or decreases by more than about 10 per cent. Several manufacturers today produce star trackers and we can recommend them to you. But remember that unless you also have a star tracker on the declination axis of your telescope, you still will not be able to leave the eyepiece for long periods of time.

Spectroscopy

The single most valuable telescope accessory for the professional astronomer is the spectrograph, and this chapter would not be complete without a few general words on the topic. At the outset let it be made clear that the relative insensitivity of the dark-adapted human eye to blue and violet light and to deep red light limits the use of a spectro*scope*. Barely more than a quarter of the 'visible light' spectrum of a star or nebula can be seen visually unless one is looking at a very bright star through a moderately large telescope. Therefore, most of what follows pertains to spectro*graphs* and what one can photograph with a modest-sized telescope and a simple spectrograph.

The primary purpose of spectroscopy is to identify the atoms which produce the gaps or bright features in the spectrum of a star, planet, comet or galaxy. Thus, one can tell the composition of the gases which may be present in celestial objects. Also, one can determine temperatures, since a high temperature causes an atom to lose some of the electrons which orbit around its nucleus and make its spectral fingerprint different. Consequently, the spectrum of a hot gas can have a radically different appearance from that of a cool one.

Find a book on spectroscopy and read more if this important branch of astronomy interests you. Any good general astronomy text will cover quite a lot of the subject, and will recommend other reading.

Buy, build, or borrow a spectroscope which you can attach to your telescope. Later convert it into a spectrograph simply by putting a piece of photographic film in the focal plane of the spectroscope (at least, in principle it is that simple; in practice design alterations will have to be made). The advantages of photography also apply to spectrography: long exposures reach fainter stars than the eye could ever see; parts of the ultraviolet and infrared spectrum can be registered; the record is permanent; and you will have become, to put it succinctly, the 'compleat' astrophotographer.

Fig. 19.1. NGC 4321, spiral Galaxy in Virgo.

Do your own thing...

CHAPTER 19

Among the pleasures that come from an involvement with astronomy and astrophotography are the endless opportunities to build or modify things. Your first project may well be the building of your own telescope (see Chapter 18).

Some very simple modifications using standard equipment can turn an ordinary camera into an automatic device. For example, an old electric clock motor can become part of an occulting shutter, permitting timed exposures of varying lengths, just as a simple kitchen timer can be made to close a camera an hour after you have opened it. The opportunities for invention are limited only by the resourcefulness of the 'tinkerer'.

The efforts of one stargazer in building a more efficient felt-lined dewguard quickly enter the common domain after a new design has been seen by other telescope-owners, say at a star party. Sometimes it goes so far that an idea featured in the 'Do it yourself' section of an astronomy publication finds its way rapidly into enterprising commercial hands, so that yesterday's unique home-built electronic tracker or camera guiding system becomes tomorrow's advertised product available to all. In this way, everyone benefits – especially those who could not have built such a system on their own.

A visit with my co-author Bill Liller to the Whipple Observatory on Mount Hopkins in Arizona yielded useful insight into the workings of the professional astronomer and his instruments. It proved that there are practical and easy-to-implement lessons which the amateur can learn from highly sophisticated instrumentation to modify and improve his own apparatus.

The first, and most striking, impression which came from merely viewing the magnificent 1.5-meter reflector was that the Mount Hopkins systems were essentially identical to those of many amateurs, with one principal difference: where the average home-built reflector at a star party may be in the 8-inch range with a price tag in the low three figures, here was a 60-inch optical giant costing several million dollars – certainly by the time the dome and the peripheral equipment were all added to it. Yet the similarities remained once I had overcome an acute sense of aperture envy. Seeing how Professor Liller focused this instrument was a revelation which I wanted to share with fellow amateurs, especially those who, like me, work in astrophotography where critical focusing can make or break any exposure, however carefully or patiently taken. Rather than turn a focusing knob or twist a dial, an electric motor was used to move the mirror in or out at the touch of a button on the hand control unit. Forward and back it went, whining a little as a first setting was tried. When the motor moved the mirror forward, a little digital readout on a console whirred forward also, presenting a four-digit number which rapidly raced from 2010 to 2056. A further correction increased the readout up to 2066, another forward touch produced the number 2073. I was asked to make a note of this number, even as Professor Liller deliberately threw the instrument completely out of focus again, to try a second focusing cycle. Starting at about 2020, the readout quickly spurted to the familiar 2060 and after several short searching motor bursts, came to rest on 2069. This number also was noted. A third focusing run, this time working up from 2000, quickly homed in on the now familiar high '60s and settled on the number 2068. Adding the three numbers, 2073 plus 2069 plus 2068, gave 6210; dividing this number by three (the number of focusing tryouts made) yielded 2070 as an average setting. This was the number onto which the telescope was then mechanically focused without further visual checking.

I learned that the four-digit number on the console was completely arbitrary. Its sole purpose was to establish a numerical frame of reference on which an optical setting selected from averaging different results could be based.

The rest was easy: what's good for a million-dollar instrument is good for any Celestron, Mead, Dynamax or similar reflecting telescope. Fig. 19.2 shows a simple manual counterpart to the costly electric motor focusing drive on Mount Hopkins. A moving pointer attached to the knurled focusing knob and a dial pirated from a broken electric timer fixed to the 'scope easily perform the functions of the electronic readout. A reverse application is possible, where the dial moves with the focusing knob and the pointer remains fixed. Since any number can serve as a starting point to obtain an average reading taken from different focusing, the 48 half-hour divisions on

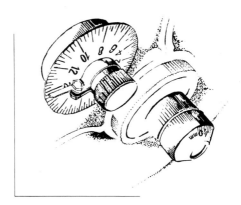

Fig. 19.2 What's good for a million-dollar instrument is good for any amateur reflecting telescope.

my dial perform extremely well. When attaching any other graduated dial to the telescope with twin-sided tape, no special position will have to be selected. To focus critically, all I do in my case is to record the hour and minute settings of three or four focusing attempts and then find the average for my optimal adjustment.

One important point is worth remembering here: the linkage between the adjusting knob and the movable optics always has some play in it which can cause focusing inaccuracies when readings are taken on alternate clockwise and counterclockwise adjustments. Therefore, after throwing the instrument out of focus between tests and when turning the knob for the finer adjustment, always approach your final setting from the same side. I turn mine clockwise – to the right – and I have had the right focus for my photographs ever since.

The problem of dew, briefly touched upon before, is of greatest importance to the observer on nights when the humidity is high and moisture settles stealthily on your equipment. Several methods are available to combat dew. One popular method is to keep an electric hairdrier handy and to direct intermittent brief bursts of hot air at the affected areas, especially lenses and eyepieces. (Do not aim the drier directly at lenses or mirrors during photography or the glowing heat-wires will leave strange patterns on your film.)

Another workable solution, for both catadioptric and refractor instruments, is to fashion a tube, matching the diameter of the telescope barrel, out of any flexible sheeting, such as corrugated cardboard or thin aluminum. This tube is press-fitted into or over the forward end of the telescope. Such a 'dew-guard' can be of any reasonable length since its purpose is merely to create an air chamber forward of the main or corrector lens. If lined on the inside with felt or other insulating material, the air within this cylinder is maintained at the temperature of the telescope usually slightly higher than that of the ambient night air surrounding the sleeve. This device will not stop the dew from settling on the lens eventually, but will retard condensation between bursts of hot air from the electric hairdrier. If used in conjunction with some permanent heat source, a heated dew-guard will keep a telescope operable in all but the highest humidities.

The creation of heat through electricity can be safe and easy when performed through the use of low-voltage bell transformers in combination with identical ceramic 'resistors' which can be obtained inexpensively from radio or television repair shops. Wiring several of the tiny oblong units in series, by attaching one to the other to form a chain, and then connecting the two extreme wire ends to the 12- or 16-volt terminals of a bell transformer will produce heat in the resistors. As a rule of thumb, six '5 watt–5 ohm' ceramic resistors attached to a 16-volt bell transformer will produce a fair amount of heat. Such a chain of resistors attached to the inside of a dew-guard tube will

Fig. 19.3. Electrical resistors in combination with a dew-guard will keep optics free of moisture when humidity is high.

produce astonishing results. (The more resistors you have, the lower the temperature; the fewer the resistors you use, the more heat you get.) Check with a radio or television service person, they should be able to assist you (see Fig. 19.3).

For other sources of heat, consider a small 15-watt light bulb covered carefully with black electrical tape, or an electric soldering iron, or even an electric heating pad. Be sure anything you use which plugs into household current outlets is approved by the relevant safety body, in safe working order, protected from moisture itself and properly grounded. Whatever the electrical device, if it can be plugged into a regular outlet it can also be timed and can be made to go on automatically as dusk settles and to switch itself off again at dawn. All that is needed is an ordinary household electric timer. Wrap timers in polyethylene bags. A photoelectric switch can also be employed. Such a switch will turn on whatever is connected to it as darkness falls and turn it off when it gets light. The time at which it turns on or off can be affected by clouds or haze and therefore such a switch will not always keep a clock-like schedule. Still, the sensors are accurate to within a half hour or so. Read the instructions and the ratings which are printed on the unit. Like light bulbs, they give wattages, i.e. capabilities to switch equipment only up to a certain rating as stated. Ask the salesperson in the hardware store, but chances are your equipment will draw very little current. It can happen that you will have to buy the next larger model if your power needs are high. Combining this little mastermind with your painted light bulb, your heating coil or your low-voltage transformer allows endless opportunities. You can protect unguarded equipment from dew in your absence and perform many other functions automatically.

Obviously with fluctuations of up to half an hour in time, and because light itself is needed to switch a photocell, these devices cannot be used to start or stop photographic exposures. For that purpose an array of electric on/off timers is available. Together with other devices, such clocks allow improvisations to open and shut cameras, to start pre-set guiding motors and to perform many other duties. Again, you can open a camera yourself and then have it closed automatically, perhaps in your absence or at some specific time before dawn – while you are in the arms of Morpheus. Available electrically controlled and motorized cameras which advance films may need a little 1.5-Volt calculator transformer 'support' instead of their basic battery. In combination with low-voltage timer systems, which turn lawn-sprinklers and similar devices on and off, and with little solenoids (see below), patrol equipment of almost professional standard can be assembled. Such systems can take up to 20 sequential photographs per night. Sprinkler timers can be adjusted to determine timing and lengths of individual exposures. Remember the hat trick? That's when we used a camera which was already held open by means of a locked cable release and the shutter set on the 'B' (bulb) setting. Needless to say, such an arrangement will work only in relative darkness during night conditions, when stray light will not leak into the lens and ruin the sensitive film. All that needs to be done to achieve a long exposure automatically is to duplicate the movement of the cardboard, but this time with a circular motion.

By rotating a so-called 'occulting shutter' (see Fig. 19.4) in front of the lens, the automatic hat trick can be readily duplicated. If an electric clock motor is used and the disc is attached to the axle normally turning the minute hand, then an opening slice 5 minutes wide will produce a 5-minute exposure. A 10-minute slice will yield 10 minutes, and so on (see Fig. 19.5).

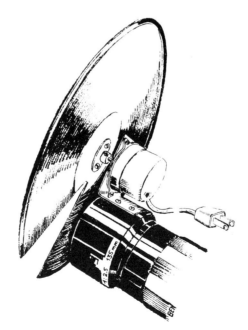

Fig. 19.4. A rotating occulting shutter permits taking sequential photographs on the same negative or color slide.

In guided photographs and by using this scheme with regular or wide-angle lenses, it is possible to take several consecutive exposures of a moving object on the same frame of film. The image of any body in motion will occur as a line of equally spaced dots where all the other stars are held in their place by the guiding motor. Only rapid movement can be detected in this way among all other objects which appear to remain stationary. All one has to do is to let the shutter revolve several times in one night after cutting one 5-minute exposure into an otherwise black disc. Phonograph records serve the purpose well and can be scored with a sharp knife and then carefully clipped with needlenose pliers. Be sure you spray the glossy surfaces with matt black paint before using that scratched record. Obviously, accurate polar alignment becomes an important consideration in such work.

An extremely simple and easy to build cable release is shown in Fig. 19.6. It could be called the proverbial 'better mousetrap'. The contraption utilizes a

Fig. 19.5. A 10-minute slice will yield a 10-minute exposure.

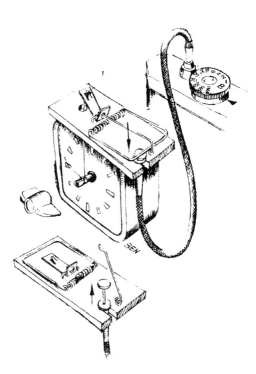

Fig. 19.6. A 'better mousetrap'. Note: this method is used to close a shutter, not to open it.

Please note: you *open* the camera, the mousetrap causes it to close.

The use of low-voltage solenoids, those magnetic devices which can be caused to push or pull through the application of electric current, open a great abundance of possibilities for the tinkering amateur. A very simple application is shown in Fig. 19.7. Here a 'Clippard' low-voltage solenoid is used in connection with an ordinary cable release. First, the small plate, to which pressure is applied with a thumb, is snipped off. Then the flange, which is held between forefinger and middle finger, is ground or filed down to fit snugly into the solenoid housing and is glued into it with epoxy or a similar strong adhesive. Activating the solenoid with the cable release in place before gluing the two together allows you to check for the best position.

The wonderful road which lies before you leading to home-built, fully automated equipment is merely suggested when one begins to think in terms of an electric timer turning on the current to the low-voltage transformer which, in turn, activates the solenoid, which instantly pushes the plunger on the cable release into the camera, thus opening the shutter. Through the setting of a suitable interval on the electric timer, any exposure from 20 minutes or so and upward can be programmed. After the selected time has elapsed, the current is turned off again and the shutter automatically closes. You may want to add the sprinkler-timer for all-night patrols.

When the above system is used with an equatorial drive, long-exposure photographs can be taken automatically by pointing the camera to the correct Right Ascension and declination several hours before the intended exposures. You may be aiming at some point which is still near or below the horizon at the time you normally go to bed and still take fine meridian photographs at 4 o'clock in the morning. Use such setups together with the heating devices suggested earlier, shroud all equipment in a polyethylene bag, and you will be set to take an entire

simple kitchen timer, as shown, or any other spring-type alarm clock which is so constructed that the wind-up stem of the alarm mechanism rotates while the alarm sounds. The kitchen timer will function with exposures of up to one hour's duration. The alarm clock approach permits much longer exposures. The mechanical function is performed by the mousetrap which can translate a minimal movement into a major motion through the use of springs.

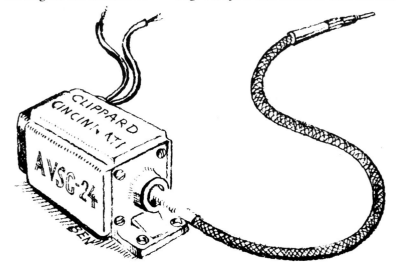

Fig. 19.7. A low-voltage solenoid, when used in combination with a sequential timing device and a motorized camera, permits many wonderful methods to control cameras and exposures. Very useful patrol and survey work can be performed with such devices in a fully automatic manner.

series of potential discovery pictures each night, many of them while you are sound asleep.

In the Rube Goldberg device which I built for automatic meteor photography, I carried the solenoid principle to its absolute extreme and used the devices not only to open and close the shutter but also to advance the film. That is how ARCTU (Automatic Recording Celestial Tracking Unit) came about (see Fig. 19.8). ARCTU, of course, was the recorder of Nova Cygni's explosive birth. With a little help from some friends, you may be able to build your own similar system. If you have access to (or adequate funds to purchase) a motorized camera, you will find that it performs almost all of the functions listed above and merely needs the addition of a timing device to do patrol work for you.

As you become the owner of a telescope – and to avoid problems afterwards – you should review the question of portability of all your equipment. Depending of course on the type of mount you have purchased and the size of the instrument and its weight, a conventional tripod carried in a car may be only one of the various ways open to you to transport your components into the field. Sturdy supports, especially for larger telescopes, can be costly but there are some exciting alternatives. If you are a city-dweller and want to travel to higher and darker ground for your observing sessions (you will, you will), it may be worth your while to check out the least expensive

Fig. 19.8. ARCTU (Automatic Recording Celestial Tracking Unit). (A spoof of the use of overblown acronyms by the scientific world.) Fully automatic system using a 35-millimeter camera activated by a mundane lawn-sprinkler timer, a low-voltage solenoid, and a powerwinder. Entire system mounted in an equatorially driven waterproof housing.

types of boat trailers. With some minimal hacksaw surgery, the long forward tongue (designed to allow for the length of the boat) can be removed and reused to build some kind of a superstructure to which the equatorial mount is then attached. To afford the stability needed (and to assist in leveling and sighting-in the system), two boat trailer jacks are added to the rear of the vehicle which, together with the conventional forward jack, permit the easy lifting of the whole framework in a stable three-point mode. I have found it useful to remove a little wheel from the two rear jacks and to use the front one to swivel the entire trailer during alignment.

A battery to drive the equatorial motor and dust-proof case in which the optics can be carried (wrapped in polyethylene) all can be contained in such a portable trailer, adding to its basic weight and stability when set up. SCAMP (Self-Contained Astronomical Mobile Platform) (Fig. 19.9) is a home-built unit assembled in this fashion. It can be easily towed by any vehicle. It still may be best to store the telescope itself in the cushioned dust-free comfort of the towing vehicle while traveling.

Creature comfort during the course of the night while observing or photographing must not be forgotten. First consider warmth, because it can get very chilly after midnight, even in summer. Warm underwear, two pairs of socks and skiing gear all add to relieve possible discomfort (see Chapter 16).

An uncomfortable position at the guiding eyepiece can turn into neckstrain agony even during a 15-minute exposure. Various solutions to this common problem offer themselves, for example, a box which you can build out of plywood for your specific needs and which affords three different seating heights depending on which way it is turned (Fig. 19.10).

Fig. 19.9. The 'gypsy' approach to astronomy: 'Have telescope, will travel'. A boat trailer converted to transport and support a 14-inch Celestron reflector. The name of the system is SCAMP (Self-Contained Astronomical Mobile Platform).

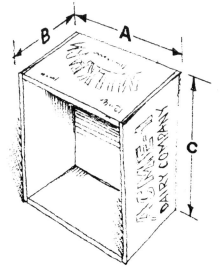

Fig. 19.10. A simple box affording three different positions for the observer at the telescope. Build your own for the best combinations.

DO YOUR OWN THING... 145

A much more ambitious scheme involves a lifting dolly either with a crank or electrified (see Fig. 19.11). Not only does such a seat permit platform adjustments to within fractions of a centimeter, but the entire hoist can also be employed to lift telescopes or other gear into position.

You will usually be confined to 12-volt battery systems when you're out in the field. Almost all telescope-related gear is available for use with 12-volt batteries as well as with household circuits. Converters can change 12-volt current into 110 or 220 volts and *vice versa*. The possibilities are endless.

Fig. 19.11. The 'electric chair': the extreme in comfort. This conversion of a hoist does double duty: it also lifts the telescope barrel of the reflector into position between the tines of the telescope fork for easiest assembly.

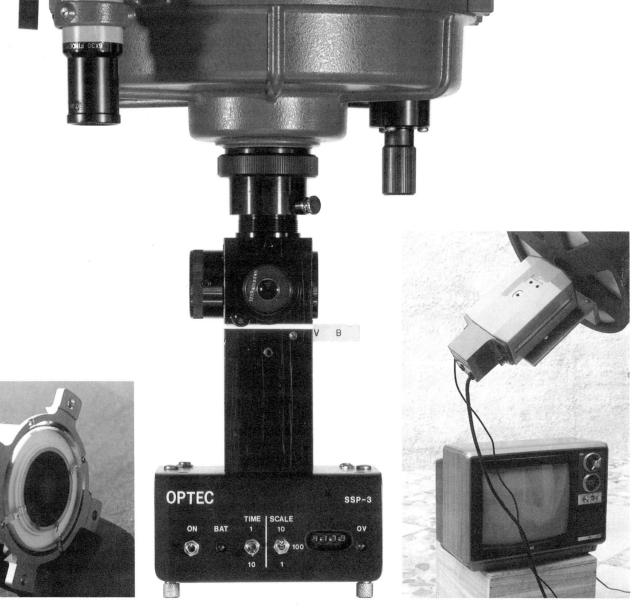

Fig. 20.1. Exposure times can be greatly reduced by using image intensifier tubes. Focused light strikes a sensitive electronic surface, and the tube then amplifies and refocuses the image.

Fig. 20.2. A photoelectric photometer, a highly sensitive light meter that one can buy (or build) for not too much money, permits the amateur to measure star brightnesses and colors with full professional accuracy.

Fig. 20.3. A low-light-level closed-circuit television camera, especially if coupled with an image intensifier tube, can make stargazing as comfortable as watching living-room TV.

Chapter 20: Closed-circuit TV and other aids

A number of years ago, a certain manufacturer of a high-quality line of refractors ran a most persuasive advertisement which showed a father and son outdoors at night, obviously engrossed in viewing some astronomical object through a telescope, while in the background you could see through the open window of a house a television set, turned on but unwatched. The caption read, 'Johnny's not watching TV tonight'. A nice idea, but why not put whatever Johnny and his dad were viewing telescopically on TV? Then the whole family could watch. Nowadays, closed-circuit TV systems are everywhere, as security aids in stores and banks, in schools as classroom aids, in factories for remote viewing. TV tape recorders will soon be as common in the home as hi-fi systems and washing machines.

This chapter is not only about TV for telescopes, but also about the use of several different kinds of commercially available devices, mainly in the category of electronics. You might like to experiment with some. There are, after all, a number of excellent electronic light detectors besides those two well-proven receptors, the eye and the photographic film.

Light meters and phototubes

It has been known for some time that certain materials spontaneously generate electricity under the action of light. If placed in a vacuum and illuminated, they immediately emit a cloud of electrons. Anything that is positively charged will attract these negative electrons at once. Given enough electrons attracted to a wire maintained at a positive voltage, a measurable current results; if the flow of electrons is not too great, it is possible to count the electrons one by one.

A vacuum tube containing only the light-sensitive surface and the collecting wire is called a *phototube*, but the more frequently used device is the *photomultiplier* (or simply PM) which does what its name suggests: each electron given off by the target plate is 'multiplied' several million times by bouncing it around in a maze of specially coated metal boxes, producing finally a more easily measured current or train of countable pulses.

Nowadays, one can buy a good PM tube mounted in a convenient little package, the *photometer*, ready to slip into the eyepiece tube of your telescope. Several companies manufacture excellent photometers and their advertisements appear in the serious amateur magazines. The price of the entire package may be comparable to what you paid for your telescope, but what you will have is a professional-quality instrument with which you can measure star brightnesses and color with the highest accuracy attainable. Anyone can truly duplicate major observatory work.

To use the photometer, one looks into the eyepiece and centers a star in a focal plane aperture small enough to block out the light of any nearby stars. One then removes (by a simple turn or pull of a knob) the small viewing mirror from the light path, thereby allowing the starlight to illuminate the PM. The output current is then read off the meter, and one has a number which is exactly proportional to the brightness of the star. If the star were to double in brightness in the next 10 minutes, so would the meter reading. To convert the readings into magnitudes, one only needs to measure, from time to time, stars of known magnitude close by in the sky. If the stars are faint, correction will have to be made for the skylight also admitted by the small photometer aperture. This is simply done by taking a reading on a star-free patch of sky and subtracting its value from the star-plus-sky reading.

In virtually every issue of the major professional astronomical journals, one can find several important scientific papers based on measurements made in precisely the manner described above. The better-equipped amateurs in the AAVSO use photometers, often built by themselves, perhaps with some advice from a friend or neighbor with a knowledge of basic electronics. The addition of a photometer and the ease and reliability of the readings it permits, can make your data most valuable to the AAVSO. Good and frequent observations can then be forwarded to the scientific community on your behalf.

Amplifying light: image tubes

Strictly speaking, light cannot be amplified, but it can be intensified. It is thus possible to produce an image of a comet, galaxy or star field which appears brighter to the eye (or photographic film) than the original. The magic device that does this job is called an *image tube* or *intensifier tube*. It operates on exactly the same principle as the phototube: light is focused onto a light-sensitive material in a vacuum, producing a cloud of electrons. By means of properly applied electric fields, the electrons are herded and refocused onto a viewing screen similar to that in a miniature TV set. The science of focusing electrons with electric fields is highly developed, thanks to the TV industry. The resulting picture on the image-tube screen is a sharply defined and intensified image of the original, that can be inspected in detail with a magnifying lens or photographed with a camera having a close-up attachment.

In principle it is possible to mate two (or more) image tubes together by interfacing screen and sensitive surface, and getting in effect a multiplied intensification. The manufacturers do this quite well by omitting the glass between the two tubes and putting the sensitive surface of tube No. 2 directly on the phosphorescent screen of tube No. 1. While a much brighter image results, some loss of image sharpness occurs. Nevertheless, two-, three-, and even four-stage image tubes are commercially available and work amazingly well.

Many of the big-name electronic tube companies make high-quality image tubes and advertise in the leading amateur magazines. For a reasonable cost, one can buy a battery-operated, one-stage tube of excellent quality with an 18-millimeter-diameter sensitive surface. Many are provided with a fiber-optic faceplate over the final viewing screen; then instead of the light from the glowing phosphor radiating out in all directions, it is channeled by thousands of glass fibers so that close-up pictures of the final image are even brighter. High-quality photographs can be taken simply by pressing bare film directly against the output end and letting the fiber optics carry the light directly to the emulsion. With a simple one-stage tube, exposure times can be reduced by as much as *ten times*; with a two- or three-stage intensifier, it is possible to take photos of some of the brighter nebulae and galaxies in seconds instead of minutes. You might keep this in mind the next time you spend several hours guiding at the eyepiece on a cold, winter's night.

Before leaving this section a few words should be added about the electronograph which, in some ways, represents the ultimate in image tubes. Here, by means of an ingenious vacuum gate, a photographic plate is actually put inside the evacuated tube and the electrons released by the sensitive surface are focused directly onto a special emulsion which is particularly sensitive to electrons. While we can hardly recommend that amateur astrophotographers undertake electronography as a hobby, at least you can appreciate its value to the professional and understand how it works.

Closed-Circuit TV

In recent years, the TV industry has developed a special camera tube which possesses high sensitivity. One advertising ploy which strikingly demonstrates this super-useful characteristic is simply an ordinary photograph of some scene, such as a simulated warehouse burglary in progress. In the foreground you can see the outlines of a TV camera aimed at the shadowy figure, but there, in the lower right corner of the picture, the guardhouse TV monitor is brightly displaying the robbery in progress.

The names applied to this particular type of camera tube are 'silicon vidicon', low light-level TV, or simply L^3 TV. By means of these recently developed and highly sensitive television tubes, one can provide surveillance of poorly illuminated areas such as dark corners of factories and parking lots at night, as well as the surface features of Saturn. Because even the best available TVs do not yet yield the ultimate in picture quality, at the present time professional astronomers use TV mostly to provide warm-room remote guiding of telescopes rather than as the last word in the recording of pictures. However, improvements come fast in money-intensive businesses like the TV industry, and we can certainly expect additional applications of TV to astronomy in the near future.

We have already become accustomed to seeing TV images of the clouds and craters on other planets and satellites. Whoever would have thought 30 years ago that one would some day sit in one's living room and see a sharply defined picture of a distant moon called Phobos, 21 kilometers in diameter, filling up the 21-inch screen of a television set with a name like Quaser.

How much can you see if you mount an L^3 closed-circuit TV on an 8-inch telescope and turn up the brightness and contrast on the monitor? The simple answer is not encouraging: not much – just yet. The Earth's Moon presents no problem and having your home TV screen filled with its craters imaged brightly and distinctly by your own telescope will impress even those who are used to seeing on their TV screen the finger signals flashed by the catcher in a baseball game. You should even have no trouble seeing the phases of Venus with your telescopic TV. But the cloud belts on Jupiter and the rings of Saturn will be a real test, and stars fainter than about fourth or fifth magnitude will be beyond reach. The reason for the disappointing performance is not that the camera tube lacks sensitivity; the trouble lies in the circuits that follow. A standard television is designed to provide a continuous barrage of images of one-thirtieth of a second duration; camera exposures teach us not to expect many stars with exposures of fractions of seconds: circuits which deliver one picture a second have been designed and work much better in nearly all astronomical applications. With proper circuitry they should have 30 times the sensitivity of standard antiburglar closed-circuit TVs. If your cousin, brother-in-law, or next-door neighbor happens to be a clever electronic technician, perhaps you can get him or her interested in the problem and have the necessary alterations made.

If you would like to try a closed-circuit TV, mount the TV camera rigidly to your telescope, with its lens approximately one centimeter behind the eyepiece. The camera lens should be opened to its widest aperture and focused at infinity. Adjust the telescope eyepiece to give a good focus for your eye. Start by aiming at the Moon and focus by racking in or out the eyepiece while watching the monitor screen. If a crystal-clear picture does not leap into view, it is probably because the brightness and contrast controls need adjustment. (Presumably you have already found the best setting of the other knobs by living room experimentation.) To reach the faintest possible magnitude limit, unscrew the TV camera lens and remove both it and the telescope eyepiece. By looking into the camera case, you should be able to see the light-sensitive faceplate of the vidicon tube located a few millimeters behind the normal position of the camera lens. That surface of the tube must be exactly in the focal plane of your telescope, so make the necessary adjustments and try again on the Moon, a planet, or a moderately bright star.

Many closed-circuit TV cameras are made to operate with any standard home TV receiver, color or black and white; the camera output lead is clipped directly to the antenna and the set tuned to Channel 3 or 6 (usually). If you want to record your efforts for posterity, buy or borrow a TV tape-recorder and plug it in.

If you want to photograph your Moon image directly from your TV screen, don't forget that to avoid uneven screen illumination the exposure time should be at least several times longer than the scan rates, one-thirtieth of a second. In fact, less 'snow' and an improved picture results if you darken the screen and close down the camera lens so that the necessary exposure time is in the vicinity of a quarter of a second. In principle, if the set is adjusted to give a nearly black sky, one can realize all the benefits of long astronomical exposures but in the warmth and comfort of the study or whatever the TV-viewing room is. And there are no commercials.

Most of the techniques described in earlier chapters can be applied to L^3 closed-circuit TV, such as the use of tele-extender or telecompressor lenses. If you really decide to go all the way and purchase a complete arsenal of detectors, a logical combination would of course be an L^3 closed-circuit TV focused on the output screen of a two- or three-stage image intensifier tube. With a proper match-up you will be able to record approximately 25 per cent of the photons that arrive on your telescope mirror and, in fact, when looking at a faint galaxy or nebula, you can actually see the individual photon events as they occur on the first sensitive surface of the image tube.

'Why', you may ask, 'isn't it possible to see 100 per cent of the photons and thereby increase four times the brightness of the final image?'. Alas, no light-sensitive surface is known which reacts to every photon which strikes it. However, 70 or 80 per cent is possible, and the most recent development in light detectors, the charge-coupled device, or CCD, does achieve this high efficiency. Basically, a CCD is an array of thousands of individual photosensitive elements whose outputs are recorded with a computer. When an exposure is completed, the computer is read out picture element by picture element. All these 'pixels', as they are called, can be combined on a video screen for viewing, or printed out using x's and o's to provide tones and half-tones. The professional wanting to measure faint magnitudes will be more interested in the individual photon counts per pixel.

In electrons, pixels and CCDs lies the future of astronomical research, not only for Earth-bound telescopes but also for those borne aloft by satellites.

Fig. 21.1. A meteor shot by during this long exposure taken by Ben Mayer. With fast film and a fast lens, several meteors can be photographed in one night at the time of a major shower. A rotating shutter in front of the lens produced the dashes and makes it possible to estimate how fast the meteor was traveling (about 45 kilometers per second for this example).

Miracle objects

CHAPTER 21

Toward the end of the eighteenth century, a Frenchman named Charles Messier brought out a catalog listing over 100 objects, many of which he had discovered or seen in the course of his dedicated search for comets. The exact number remains under debate because of later additions and deletions.

Messier was comet crazy – rather than star struck – and one of the reasons he published his now famous catalog was to keep from wasting time looking at non-cometary objects. He carefully numbered these after listing their exact location with coordinates. His catalog includes brief descriptions of what he had observed through relatively primitive optics. Messier had moved from Lorraine to Paris and became assistant to the astronomer J.N. Delisle. Before long, the young man had come to run the observatory in which, at first, he had been merely apprenticed. In time he succeeded his mentor. It is known that this passionate comet-seeker did his observing from an octagonal tower of the Hotel Cluny in the Latin Quarter of Paris. The Cluny was not a Hilton or Sheraton type of establishment, but rather a famous and splendid fifteenth-century residence which, like other similar palatial houses in France, was called an 'hotel'. Fewer than 50 comets were known when Messier embarked on his search and in time he was to add 21 discoveries of his own. Of these, 15 had never been observed before.

Since the terrestrial latitude of Paris is approximately 49 degrees north, the items which Messier listed are confined to the Northern Celestial Hemisphere, ranging from the most northerly No. 82 ($+69°\ 56'$) in the constellation of Ursa Major down to No. 69 ($-32°\ 23'$) in Sagittarius. All the Frenchman knew as he sat in his tower and followed the orbital movements of the comets known at the time, and searched for new ones, was that many of the objects he had spotted simply would not budge from their positions. Obviously, then, these celestial features were not comets which would add to his impressive list of discoveries and carry his name into posterity.

What the conscientious Charles Messier could not know, nor ever suspect as he recorded his findings with a quill by lantern light, was that among the 100-odd disappointing non-comets, were all the major northern celestial attractions; these nebulae, clusters and galaxies would immortalize the name of Messier and be linked for all time with the man and his work which extended from the year 1754 until he died in 1817 at the age of 86. His eternal reward will lie in the initial 'M' which today precedes his catalog numbers and which stands for the name Messier.

It is worth noting that the telescopes Messier used had speculum mirrors made of polished metal alloys. As far as is known, he most frequently used an instrument with a $3\frac{1}{2}$-inch aperture until he was put in charge of the main instrument of the observatory, which was Deslisle's own 8-inch Newtonian. Thus, whatever Messier saw can easily be seen with any modern amateur telescope.

To me the 'M' stands for more than just the first initial of the Parisian 'comet ferret', the nickname which King Louis gave his astronomer. 'M' can also spell 'magic'. Just to mention some of the better known magic objects, here are their numbers and names: M1 Crab Nebula; M8 Lagoon Nebula; M13 Hercules Cluster; M17 Omega or Horseshoe Nebula; M20 Trifid Nebula; M27 Dumbbell Nebula; M31 Great Andromeda Galaxy; M42 Great Nebula in Orion; M44 Praesepe or Beehive; M45 Pleiades; M51 Whirlpool Galaxy; M57 Ring Nebula; M64 Blackeye or Abalone Galaxy; M97 Owl Nebula; M104 Sombrero Galaxy. This will make it possible to start an observing or photography program during any given month of the year and permit working one's way around the 24-hour celestial sphere. There should be at least 3 or 4 hours of good overhead (meridian) observing during five or six moonless nights every month and thus there should always be several Messier objects ready and waiting for you.

By going out and viewing or shooting soon after darkness falls, it is still possible to catch objects about 'to set' in the west. Similarly, by staying up extra late – into the hours past midnight – one can catch objects on the ascendancy. By combining such early or late observations or photography, it is possible to compensate for anticipated absences when one cannot shoot or to make up for pictures missed, perhaps during a vacation month in summer. Nevertheless, as stated earlier, it is always best to aim for photographs when objects are at their highest in the sky so that their light need pass through only a minimum of atmospheric pollution.

Although his catalog is probably the oldest list of its kind still in use, Messier's was not the only one to be published. There are many other such registers. One in particular must be mentioned: it is the 'New' General Catalogue usually referred to as the NGC. This compendium of nebulae and clusters of stars, by John L.E. Dreyer and published by the Royal Astronomical Society of England in 1888, lists 8000 items, among them all but two of the Messier objects. Thus it is that the Owl Nebula is known as both M97 and NGC 3587, and the Andromeda Galaxy as M31 and NGC 224, to give just two examples.

Today we know much more about the type of objects which Messier and Dreyer first located, charted and numbered. The early catalogs lumped together globular clusters, those ball-shaped aggregations of stars, with nebulae, the mysterious and vast glowing dust clouds often surrounding bright stars. No differentiation was made between open star clusters and globular star clusters. The former are large groupings of stars loosely spaced which, unmagnified, appear to the eye as mere blurs of light. The globular clusters, on the other hand, consist of hundreds of thousands of stars concentrated into tight round ball-shapes, seeming almost starlike to the unaided eye but resolving into dazzling spheres of countless sparkling stars when viewed with larger telescopes.

Nebulae were also randomly tossed together in the catalogs, although we know today that many of these are galaxies. Nebulae too fall into several distinct categories, such as the diffuse galactic nebulae named after our own galactic belt, near which they are usually found. Then there are planetary nebulae which are said to resemble planets with their disc-like circular shapes but have nothing to do with the heavenly wanderers. Another type of nebula is M1, the Crab, also known as NGC 1952. This relatively new object is considered to be the debris left over from an exploding star, a supernova which, Chinese records show, occurred in 1054 AD.

Finally, there are galaxies of many different configurations, elliptical, spiral and irregular. These are also known as 'external galaxies' because they are neighboring systems looking much like our own Milky Way Galaxy. These external stellar systems may be millions, hundreds of millions, or even thousands of millions of light years away. Powerful telescopes reveal vast numbers of such groupings of stars, similar to our own Galaxy, which itself is estimated to contain some 100 000 000 000 stars.

The majority of objects, of course, were way beyond the reach of Messier's instruments and were not seen and numbered until much later, in the NGC which also employed photography for its compilation.

Long before the Messier visual catalog or Dreyer's photographic NGC, Man was preoccupied with relating the stars to terrestrial events. Chinese astronomers long ago foretold the future through comets, eclipses or new stars (novae). All these phenomena were regarded as omens pointing to occurrences lying ahead. Predictions were made concerning the births or deaths of kings, princes or other important personages. If there was a delay between the appearance of a celestial sign and the fulfilment of some momentous prediction, time usually came to the rescue: even if it took 3 years for a noted ruler to die or for the prince to be born, as long as the threat or the promise was realized, the stars could always be proven right. Prognostications depended on the knowledge of the position of some of the stars which were therefore carefully recorded, first in star maps carved in stone and later on papyri.

Even though Babylonian texts dating as far back as 700 BC describe astronomical happenings, the Greek astronomer Hipparchus compiled the first known catalog, giving the coordinates of about 1000 fixed stars, in 135 BC. Records and chronicles were also kept in Arab lands and in far-eastern regions. With the invention of printing, star charts became popular and were widely circulated.

Today, photographic sky atlases are readily available and, as outlined in Chapter 13, we can easily assemble such a photographic jigsaw puzzle ourselves, with photographs from a simple camera. Not only will the sense of accomplishment be rewarding but many of the magic M objects will show up, however faintly, and will point to areas deserving a closer look. This is where a telescope will really come into its own, because all of a sudden you will want to view treasures which are no larger than a fraction of a degree or so in diameter.

Since it is easiest to start with the bigger and brighter objects, all that remains is to establish what we can shoot on a given night. To determine which part of the sky will be overhead at any time of the year, there is a simple formula: take the number of the month, say June (sixth month) and add 3 to it. Next multiply the product by 2: $6 + 3 = 9 \times 2 = 18$. This will give you the sidereal time (R.A. of the Meridian) which will be overhead at midnight on the twenty-first day of the month in question. To be really scientific about it, here is a formula: (no. of month + 3) × 2 = sidereal time. If this figure exceeds 24, as it will in October, November and December, deduct 24 from the end result. What will emerge here is that on January 21, R.A. 8^h will be straight overhead at midnight, in February it will be R.A. 10^h, in March 12^h and so on; a 2-hour change in Right Ascension per month.

If you get a camera drive or telescope for the holidays and you go out on the night of December 25 with your new toy, here is what you will find: $(12 + 3) \times 2 = 30$, minus $24 =$ R.A. 6 hours. Now it so happens that the Orion Nebula (M42 or NGC 1976) lies at R.A. 5 hours 33 minutes and at a declination of -5 degrees in (where else) the constellation of Orion. This tells us that, weather permitting, this striking object can be seen about halfway up in the eastern sky at about 9 p.m. At midnight it will be directly overhead, barely east of the Meridian.

We might as well start in January assuming, of course, that you have aligned your instrument and set the polar axis reasonably well on the North Celestial Pole.

Let us pick one interesting constellation for each month and within each an area which will become our principal interest. The 12 objects listed below will be visible during the entire month in question, some earlier in the night sky, others a little later on. An August object will be visible past midnight in July or in the early evening hours as late as October. It should not take you long to get the hang of it!

January THE BEEHIVE
R.A. 08h 37m, Dec. +20° 09'; M44, NGC 2632

This open cluster is large when compared to other such galactic clusters. The unaided eye sees only a fuzzy patch, yet the aptly named Beehive swarms with stars when viewed through a telescope. Unless low magnification is used, it will overflow the field of even a modest instrument. Another name for the object is Praesepe which, being Latin, sounds much more impressive.

About 30 stars are readily visible through binoculars. These seemed to the ancient Greeks to be a 'small cloud', which is how Hipparchus recorded them two millenia ago. Galileo also makes reference to this grouping in the constellation of Cancer, the crab. Through his telescope he could resolve individual stars for the first time.

As distances to such open clusters go, the Beehive is a relatively close 500 light years away and fills an irregular circle of 1.5 degrees diameter.

February TWO GALAXIES IN URSA MAJOR

R.A. 09h 52m, Dec. +69° 18'; M81, NGC 3031 (*below, left*)

R.A. 09h 52m, Dec. +69° 58'; M82, NGC 3034 (*below, right*)

There is much more to the constellation of Ursa Major than meets the unaided eye. When we train our camera or telescope on the area above and to the west of the 'Big Dipper', we find not one but two splendid galaxies. M81 is in the form of a pin-wheel spiral which has been compared to the Andromeda Galaxy (see September). M82 has the appearance of an elongated patch which has recently been found to be an irregular galaxy with an explosive past. Its turbulence is believed to affect the neighboring M81.

Since both M81 and M82 are estimated to be about 10 million light years away from us, any cataclysm which you witness tonight will have taken place 10 million years ago. At such distances the two galaxies, which from our vantage point are separated by a mere 40 minutes of arc, are still over 10 000 light years apart.

The magnitudes of the two objects are 7 and 8, respectively.

March SOMBRERO GALAXY
R.A. 12h 37m, Dec. −11° 21'; M104, NGC 4594

This Messier object was not discovered by Charles Messier at all, but by his friend and colleague Pierre Mechain, another successful French comet-seeker. About 27 of Mechain's discoveries found their way into Messier's catalog. There seems to have been no professional jealousy between the two astronomers who collaborated harmoniously.

Large observatory photographs of M104 show an edge-on view of a spiral which some have said appears like a Mexican hat with a wide brim. The central lens of the galaxy is surrounded by a shadowy disc of obscure star clouds. With its symmetrical elongated shape, the galaxy reminds one of the planet Saturn with rings.

This celestial 'frisbee' is located in the southern part of the constellation Virgo and has a magnitude of 8. It is estimated to be a staggering 20 million light years away from us and measures 7 minutes of arc in width and 2 minutes of arc in apparent height. Not far from M104 lies the famous Virgo cluster where a telescope of 8-inch diameter can spot up to 100 of a total of thousands of distant galaxies.

April WHIRLPOOL GALAXY

R.A. 13^h 27^m, Dec. +47° 27'; M51, NGC 5194

The curving arms of the dynamic spiral surrounding the Whirlpool Galaxy can be followed for about one and a half turns. They seem to wind clockwise away from the center of the main galaxy, with the northernmost outer spiral appearing to be connected to a smaller, less structured galaxy.

Messier did not give a number to this second related object, but Dreyer in his 'New General Catalogue' listed it separately as No. 5195. Located in the constellation Canis Venatici, the Whirlpool Galaxy is a difficult visual object with a magnitude of 10. The angular sizes of the pinwheels are 12 minutes of arc by 6 minutes of arc when viewed together. The two connected galaxies offer a worthwhile photographic target for the advanced photographer.

May GREAT HERCULES CLUSTER
R.A. 16h 40m, Dec. +36° 33'; M13, NGC 6205

The famous Hercules Cluster, faintly visible to the unaided eye as a fuzzy star of approximately fourth magnitude, is one of the most spectacular objects in the sky. It is the largest globular cluster in the Northern Celestial Hemisphere and was discovered by Edmond Halley (of comet fame) as far back as 1714, approximately 50 years before Messier observed it as a 'nebula without stars'.

Today even modest telescopes reveal that the object is a vast conglomeration of countless stars.

Where the English astronomer William Herschel originally estimated that M13 contained some 14 000 stars, modern estimates put the number closer to half a million or even more. Observing the cluster with instruments of different apertures, from binoculars to large telescopes, increases not only the resolution of the many suns but also the understanding of the importance of superior optical aids.

Progressively longer photographic exposures yield larger and more impressive pictures until the center 'burns out' completely with the myriad stars grouped closely together in the core of M13.

June LAGOON NEBULA

R.A. 18h 01m, Dec. +24° 23'; M8, NGC 6523

The word 'June' is often made to rhyme with 'Moon'. If poets had telescopes, they would have made it rhyme with 'lagoon'. The reason is that in the sixth month this large and most spectacular nebula graces the sky. The Lagoon Nebula can never be said to be 'high in the sky' for northern observers, because it lies at −24 degrees in the southern constellation of Sagittarius. For observers in parts of South America, Australia or South Africa the lagoon passes nearly overhead.

Discovery of this splendid birthplace of new stars is credited to the Dutch astronomer Flamsteed and dates back to the seventeenth century. As in the Orion Nebula (see December) gaseous material in the region is heavily laced with hydrogen which shows up pink in color photographs and creates stunning effects.

M8 is 9 × 40 minutes of arc and is 2500 light years away from us. A sparkling open cluster lies nearby.

July RING NEBULA
R.A. 18ʰ 52ᵐ, Dec. +32° 58'; M57, NGC 6720

Like a tiny smoke ring, the Ring Nebula hangs suspended in the constellation Lyra, one-third of the way along the line connecting the stars Beta Lyrae and Gamma Lyrae. This shell of luminous gas which surrounds a hot blue, fifteenth-magnitude star is called a 'planetary nebula'. 1779 was the year when Messier first spotted this remarkable object. The faint central star's energy excites the gas in the ring, making it glow with an eerie luminescence.

M57 is about 2200 light years away from us and has a magnitude of 9.3. The longer dimension of the ring is 83 seconds of arc and its width is one minute of arc. The object is visible through telescopes with medium apertures, but the central star can only be spotted with larger instruments.

The term 'planetary nebula' is misleading. These shells of gas have nothing to do with planets at all. Today we know some 1000 such nebulae in the sky. They are formed when the outer shells of stars explode away.

August NORTH AMERICAN NEBULA
R.A. 20^h 57^m, Dec. +44°8'; NGC 7000

The North American Nebula is not an 'M' object for the readily apparent reason that it was too diffuse and faint to be seen by Charles Messier. Measuring a full 2 degrees in width and 1.5 degrees in height, this nebula could hardly have been mistaken for a comet. It cannot be said to have a 'magnitude' of itself, covering as it does a large area and having low surface brightness. The North American Nebula is not even readily visible through a telescope but comes easily into view when photographed with the simplest cameras.

A 15-minute exposure on color film with a tracking camera through a 50-millimeter lens will bring forth the unmistakable object which bears an unmistakable resemblance to the North American Continent, complete with Pacific and Atlantic seaboards, even showing a peninsula closely resembling Florida.

NGC 7000 is a vast cloud of light-emitting gas surrounded by dark absorbing clouds of interstellar dust. It is situated in the constellation of Cygnus.

September GREAT ANDROMEDA GALAXY
R.A. 0h 40m, Dec. +41° 00'; M31, NGC 224

M31, the mysterious object in the constellation Andromeda, is referred to as the Andromeda Nebula or Andromeda Galaxy. This dates back to the time when it was not quite clear what in heaven it really was. At first it was believed that M31 was a tenuous cloud of gas, possibly within our own Milky Way Galaxy. The work of the American Astronomer Edwin Hubble established in 1925 that what can be seen with the naked eye as an oblong blur is a galaxy of millions of stars, outside the Milky Way.

Only camera records can do the galaxy justice. While the nucleus is always apparent, the length of the exposure determines how much of this brightest of visible spiral galaxies we can see. Much of the structure remains hidden because it is inclined at an angle of about 16 degrees. Today we know that the north-east edge is towards us. The width of the oval we see is generally given as 2.5 degrees and the visible height as two-thirds of a degree. Detailed measurements indicate the Andromeda Galaxy to be at least 4.5 degrees wide on its long axis. Both in structure and dimension it resembles our Milky Way.

M31 is an independent stellar system 2 200 000 light years out. Two companion elliptical galaxies, M32 and NGC 205, will also be visible in most pictures taken of the region.

October SPIRAL IN TRIANGULUM

R.A. 01h 31m, Dec. +30° 24'; M33, NGC 598

The Spiral in Triangulum, the Andromeda Galaxy and our own Milky Way all belong to 'our local group of galaxies', making them neighbors in space. Even though M 33 lies some 2 000 000 light years from us, it can be considered to be close by when compared with other galaxies hundreds or even thousands of times more distant.

The magnitude of the Spiral is 6, but the object is diffuse and not easy to spot visually. Again, photography is the best way to capture its light. It will show favorably when recorded with a 135-millimeter f/2.5 lens in a long tracked exposure. Guided photography through a telescope will be rewarding.

The Spiral Galaxy in Triangulum is an ideal target for Schmidt cameras in which it will fill the field of view. Spiral arms will register in shades of blue and condensations of hydrogen will glow in fiery pinks.

November PLEIADES
R.A. 03h 44m, Dec. +23° 58'; M45

Here is the most eye-catching of all the galactic clusters and it does not even have an NGC number. An easy naked-eye object, the Pleiades allow the photographer to delve deeper and deeper into what appears to the unaided eye as six or seven stars in the shape of a tiny dipper.

Telescopes will show up to 50 stars in the region and photographs will display many more. There is a nebulous background in the Pleiades which will record well on long exposures and show a haze surrounding the more luminous stars.

M45 is an easy object for even the simplest equipment and will be a rewarding test target for observation and photography. The grouping occupies a field having a diameter of over 3 degrees. The brightest star, Alcyone, has a magnitude of 4.2. The distance to the Pleiades is estimated to be 430 light years.

This conspicuous object in the constellation Taurus is also known as 'the seven sisters'.

December GREAT ORION NEBULA
R.A. 05h 33m, Dec. −5° 25'; M42, NGC 1976

The distinctive 'belt' in Orion, those three stars familiar to many since childhood, are beautiful in themselves. Photographing the general area around them will reveal an abundance of gaseous nebulae. Still, it is the 'dagger' I see as the most rewarding target in the constellation of Orion. Specifically, the blurred object at its northern end.

Train any optical aid – from binoculars to telescopes – on this area and the most remarkable sight will unfold before your eyes: a vast cloud of gas and celestial dust, made luminous by neighboring hot stars, glows mysteriously some 2000 light years away. Almost any telescope will visually resolve a group of four stars within the nebula. These are called the 'Trapezium' because of their arrangement. They range in magnitude from 5.4 to 7.9.

Color film will yield additional wonders with transparencies showing delicate tones of pinks and blues in this birthplace of stars. The nebula measures 66 × 60 minutes of arc. Hardly anywhere in the sky is there a square degree containing as much interest and beauty.

Fig. 22.1. The prominent star just to the left of center is currently the most distant object we know of. At eighteenth magnitude this star is actually a quasar, PKS 2000–330, and the two much fainter stellar objects just above and below the quasar may be companions or ejected material. This picture was taken by this author and the Chilean astronomer Gonzalo Alcaíno, using a CCD (Charge-Coupled Device) on the Danish 1.5-meter telescope at the European Southern Observatory. The 5-minute exposure reaches to twenty-third magnitude.

CHAPTER 22

Space for amateurs

'To describe the Universe in terms of our present knowledge is to say almost nothing about everything'. With this pessimistic observation by E.L. Schucking, philosopher of science and cosmologist, we begin the final chapter which will be a kind of Astro-Digest for those who want to know the current thinking on where we stand or, more accurately, where we spin.

We shouldn't limit this book to the one corner of the Universe where we live. While many of the nearer galaxies are bright enough to be photographed even with a simple camera on a motor-driven equatorial mounting, the most distant of these are a 'mere' few million light years away. A few quasars, those stellar-appearing objects thought to be highly compact conglomerates of stars and gas approximately 100 times more luminous than the most luminous galaxies, reach eleventh or twelfth magnitude; the most distant quasars are believed to lie close to the edge of our visible universe. What about the outer reaches of this collection of everything that we know of? Even though only a few of the very best amateur astrophotographers can penetrate that far out, all of us certainly want to know what goes on out there . . . and beyond.

It should be clear to us all that although the watery sphere on which we live seems enormous, it is the merest speck in the Universe as a whole. Consider the following example. Imagine the Earth to be shrunk down to the size of a pinhead; then the size of the *visible* Universe would be roughly 15 000 000 000 000 kilometers in diameter, already an incomprehensibly large measure. If we bring back the Earth to its real size, the distance to the edge of the visibility becomes 150 000 000 000 000 000 000 000 kilometers, give or take a few sextillions.

There are also beautiful analogies when we look into the world of the microcosm. We can think of the Solar System as a giant nine-electron atom with planetary electrons revolving about their nuclear Sun. The atomic Solar System, it turns out, has orbits which are at roughly the right scale: if the hydrogen atom, with its one electron revolving around one proton, were magnified to make the electron Earth-sized, the diameter of the electron's smallest possible orbit would be just slightly smaller than the Earth's.

Earlier we spoke of the distance to the edge of the visible Universe, but what about the invisible Universe beyond? Here even cosmologists are in the dark. We may never know.

No matter in which direction we look, we find immensely distant galaxies and quasars. The distance to the remotest quasar as of today, PKS 2000–330, is some 14 300 000 000 light years. Astronomers believe we are seeing almost to the edge of the Universe.

What will we see when we build a telescope large enough to look beyond the edge? Perhaps not much more because it is possible that the combined gravitational forces of everything inside the Universe combine to prevent anything, including light, from leaving it. If there are other universes beyond ours, this same light-tight situation could well exist, limiting further visual exploration forever. Thus, it may be that even if there are other universes we will never be able to see them – unless they collide or interact in some way with us. To see 95 per cent of the way to the Universe's edge, as we think we can now do, is to reach tantalizingly far out. As we get closer to the edge, the urge will increase to reach farther yet. And anyway, it would be nice to get a better look at PKS 2000–330 and its neighbors, especially its more distant neighbors.

To us earthlings, all the galaxies appear to recede from us, and at speeds proportional to their distances. The giant irregular galaxy M 82 is 10 million light years from us, and the 'red shift' of its spectrum tells us that its velocity of recession is 400 kilometers per second. The most distant 'normal' galaxy for which a distance has been derived, 3C 324, appears to move away from us at an incredible 168 000 kilometers per second, more than half the speed of light. From that velocity, we calculate it to be about 8000 million light years away.

The evidence is that all galaxies recede from us with velocities that average around 15, 20 or 30 kilometers per second for every million light years of distance. The number is still uncertain. Quasars, which appear to be highly luminous compact galaxies in the process of formation, seem to share in this universal outward rush. Since quasars have intrinsic luminosities about 100 times brighter than that of ordinary galaxies, we can see them much farther away, which explains why the most distant object yet observed is a quasar. Its velocity of recession is an almost incredible 285 000 kilometers per second, which tells us if we use 20 kilometers per second per million light years scale factor, its distance is 14 300 000 000 light years. That enormous velocity is about 95 per cent the speed of light and since (we believe) nothing can move faster than the speed of light, we are led to the conclusion that nothing in *our* Universe can be farther away than about 15 billion light years.

Before we ask what this tells us about how the Universe behaves as a whole, there are a few questions sceptics have raised which should be recognized before we go on. In the first place, if we don't really know what quasars are, how can we be sure that they share in the distance–velocity rule of 15 or 30 kilometers per second per million light years? Some astronomers believe that the apparent recessional velocities of quasars are exactly that – apparent and not real at all. Another question is: 'Does the distance–red shift rule persist unchanged right out to the edge of our Universe?'. Couldn't it be that the fundamental laws of physics are different out there as compared to here on Earth? If so, we would have to know how the laws changed before we could draw any firm conclusions about the edge of the Universe.

As yet we have no reason to believe that any of these assumptions on which we base our knowledge of the outer parts of the Universe are incorrect. Many tests have been made and alternative theories proposed, but the basic ideas continue to hold: we on Earth, located somewhere in the Universe, share in a great outer expansion from some point not yet located. What lies beyond, nobody knows and nobody may ever know. Not a very complete picture but one we must live with . . . and can dream on.

How did it get this way? When did it all start? What set all the quasars and galaxies in motion to begin with? The outward velocities, especially their proportionality to distance, have all the earmarks of an explosion. The parts of the initial mass that now are located farthest from the center got there because they have been traveling fastest. In outer space there is little to slow down the outward rush; the pieces of Universe are propeled through what is surely close to a perfect vacuum and therefore encounter nothing to impede their motion. So the velocities we see could be the initial velocities – except for small changes caused by the forces of gravity exerted by every single particle in our Universe on every other particle. There seems to be a slight slowing down going on, so slight that it has not been detected with certainty. But it does exist, or rather, it must exist if the laws of gravity hold everywhere in the same way. As slight as the slowing down is, there is now another important question raised. Will all this outward movement eventually come to a gradual halt to be followed by an inward falling? Will our expanding Universe someday become a contracting Universe with the inward velocities slowly picking up speed until everything crushes together at the center of it all? Then what? Will this, the most fantastic crunch imaginable, be followed by a new explosion that begins the scenario all over again?

We speak of the Big Bang Universe, which originally was so named to differentiate it from an alternate hypothesis, the Steady State Universe where, gently and quietly, atoms were spontaneously created and, to make room for them, the Universe obliged by expanding. Lately, however, the Steady State Theory has fallen into disfavor, and now we ask if we are in a Big Bang Universe or a Bang-Bang Universe where the primordial explosion occurs over and over again.

It is possible to have all the galaxy and quasar outward motions slowing down (due to their mutual gravity) and never actually coming to a complete halt. In this case, we say the Universe is open; if the expansion is eventually halted, then we are in a closed Universe. An open Universe means that it can increase in size forever, that it will become infinitely large after an infinite period of time has passed by. If there are other universes out there, we would someday begin to interact with one or more. In fact, if our Universe is open and other universes do exist, we might expect at least one of these universes to be so old already that it has become large enough to pervade our own and most other universes. Seeing no sign of such intrusion, we are led to believe that if our Universe is open, we are it; there are no other universes. Then space beyond our own Universe must be completely empty and infinite. On the other hand, if we are in a closed Universe, we may not be alone; there could be an infinite number of other universes.

What about the effects of gravity on light? Any photon of light leaving the outermost quasar of our Universe heading outwards will be subjected to the combined gravities of everything within the Universe. Will it be able to escape from the Universe or not? If it can, we must conclude that the grand total (all the bits of energy, large and small, within the Universe) is *not* conserved since light is one form of energy. Given enough time, our Universe will run short of energy, producing a universal energy crisis, because so many of the different types of energetic activities at one time or another result in a spark, a glow, or a flash of light. The emitted photons could then escape from the Universe forever. Ultimately the Universe may 'run down', ending as a cold massive hulk back at the center where it all started.

If light cannot escape from the Universe, we would have a Cosmos lasting in its present energetic form forever. The Universe might change its appearance over long periods of time, and the atoms in it might alter their form and distribution, but the total amount of energy within out Universe would stay the same.

Remember that every single piece of matter in the Universe – this book, the chair you sit in, the molecules you breathe, the Earth you ride on – is equivalent to energy. By 'equivalent', we mean 'changeable into'. Suddenly changing one page of this book completely into energy would produce a blast equal to that of 20 000 tonnes of TNT, or 30 million kilowatt-hours of electricity, according to Einstein's formula, $E = mc^2$. The trick, of course, is how to make the conversion from mass to energy. Setting fire to paper only reshuffles the atoms and molecules, providing us with a little energy – heat and light. That is a chemical process. The Universe can be thought of as containing a certain amount of total energy, much of it in the form of mass. Whether some of this energy is escaping from the Universe's gravitational pull, we do not know.

Before delving into questions about how the Universe got here in the first place, we should ask how long has it been since the Big Bang – or the last Bang if there have been many Bangs. There is one way of answering this question directly. If somewhere there exists a quasar traveling away from us at the speed of light and which is now 15 billion light years from us, we conclude that 15 billion years ago that quasar was right here next to us. (We are assuming that the scale factor is 20 kilometers per second per million light years.) It happens that the universality of the distance–velocity law predicts the same starting time for everything. Fifteen billion years ago everything was together in one big chunk. It is encouraging to find that the giant globular clusters in our own Galaxy, the oldest conglomerations that we know of, seem to be about the same age too. Geologists tell us that the oldest rocks on Earth are 'merely' 3 or 4 billion years old. The Universe had already been around for about 12 billion years before this youngster planet of ours put its own sphere together.

We should emphasize that the light we are now seeing here on Earth from PKS 2000–330 has been traveling through space for 14.3 billion years. This means that today we are seeing PKS 2000–330 the way it was 14.3 billion years ago. The Universe is, in effect, a giant time-machine which allows us to look backwards in time and see things the way they were. As far as we can tell, PKS 2000–330 looks like many other compact clusters of stars and gas. The atoms seem to behave in the same way they do here and now on Earth; the light we receive appears to be no different from that which we get from the stars and gas in our own galaxy. What we are witnessing in PKS 2000–330 is a massive system, probably more massive than the average galaxy, in its earliest stages. It presumably is not much older than a half billion years – a mere child of a galaxy. Perhaps that is what our own Milky Way Galaxy looked like then.

We have said *nothing* about relativity. Einstein's famous theory applied to all of the above gives us somewhat different answers, but the fundamental ideas remain the same. If you are intrigued by relativity, good! But we will leave the intricacies of its explanation to others and wish you all the best.

How did it all start? No one knows. No new clues have been picked up in the spectrum of distant quasars. The Expanding Universe leads us to no conclusions. If we live in a Bang-Bang Universe, there is no reason to believe that there have only been a few Bangs. Many would say that the Bangs must have been going on forever, that there have already been an infinity of Bangs and there never was a start or a beginning at some distant point in time. If, on the other hand, we are witnessing, and are very much a part of, the only Bang there ever was, or the first of many to come, then we get into questions about the existence of an Almighty, a Supreme Being, or a Great Force that one day, long ago, decided that it would be a Nice Thing or perhaps merely an Amusement to create a Universe.

With that weighty thought, expressed rather flippantly, we come to the end of the book.

To the many readers who for the first time may try to capture their own views of the heavens with pictures that turn out a little blurred or trailed, we hope that soon everything will be in good, sharp focus. To many, we are sure that astrophotography will become a passion: looking for the new, the unexpected; contributing to science. We hope that for all, the skies will become a source of delight and will yield immense and intense pleasures. Our advice is: Go to it! With naked eye, with simple cameras, with lovingly crafted home-made Newtonians, or with chrome-fitted store-bought Schmidt–Cassegrains equipped with expensive closed-circuit vidicons, enjoy the Universe. To the authors of this book, astronomy is a real turn-on; may it be one for you, too!

Appendix

POLAR ALIGNMENT – THREE EASY METHODS

Method 1 (quick and dirty): Aim the polar axis of your telescope mount at Polaris (or Sigma Octantis if you live in the Southern Hemisphere) as best you can. A store-bought mount may provide and easy way to do this. Should be sufficiently good for binoculars or a wide-field telescope.

Method 2 (works well if done with care): Set the angle that the polar axis makes with the horizontal equal to your latitude. Then rotate the telescope base around the vertical until the polar axis points *due north* (or due south if you live down under).

Method 3 (can give perfect results): Do (1) or (2) above. Then center your telescope on a star that is on, or slightly east of, the Celestial Meridian and close to the Celescial Equator. After 5 or 10 minutes, note in which direction you have to move the telescope to re-center the star. If you had to move *north*, rotate the telescope base *clockwise* around the vertical by a small amount; if you had to move the telescope south, rotate counter-clockwise. Repeat the process until the telescope tracks meridian stars perfectly. The same procedure works in both the Hemispheres.

Now point your telescope at an equatorial star low in the west. If again the telescope has to be moved *north* to re-center the star, *increase* slightly the angle that the polar axis makes with the horizontal; if you had to move south, decrease the angle. Repeat the process until your telescope tracks perfectly stars low in the west – and all other stars. (The same instructions will hold for the Southern Hemisphere if you observe an equatorial star low in the east.)

Important note: Remember that a compass needle does not ordinarily point due north and south. However, when Mira (Omicron Ceti) or Arcturus (Alpha Bootis) is on the Meridian, Polaris is too and therefore will be due north. (The same goes for Sigma Octantis when Delta Velorum or Beta Pavonis is on the Meridian.)

GREEK ALPHABET

A	α	alpha (a)	N	ν	nu (n)
B	β	beta (b)	Ξ	ξ	xi (x)
Γ	γ	gamma (g)	O	o	omicron (o)
Δ	δ	delta (d)	Π	π	pi (p)
E	ε	epsilon (e)	P	ρ	rho (r)
Z	ζ	zeta (z)	Σ	σ	(ς final or C c) sigma (s)
H	η	eta (ē)	T	τ	tau (t)
Θ	θ	theta (th)	Υ	υ	upsilon (u)
I	ι	iota (i)	Φ	φ	phi (ph)
K	κ	kappa (k)	X	χ	chi (ch)
Λ	λ	lambda (l)	Ψ	ψ	psi (ps)
M	μ	mu (m)	ω	ω	omega (ō) (or Ω)

BIBLIOGRAPHY AND FURTHER READING

Abell, George O., *Exploration of the Universe*, 4th edition. Philadelphia: Saunders College Publishing, 1982.

Briggs, G. A. and Taylor, F. W. *The Cambridge Photographic Atlas of the Planets*, Cambridge, UK and New York: CUP, 1982.

Burnham, Robert, Jr., *Burnham's Celestial Handbook*, 3 volumes, New York: Dover Publications, 1978.

Kukarkin, B. V. and others, *Catalogues of Variable Stars*, 3 volumes and supplement. Moscow: USSR National Academy of Sciences, 1976.

Kunitzsch, Paul, *Arabische Sternnamen in Europa*. Wiesbaden, Federal Republic of Germany: Otto Harrassowitz, 1959.

Mallas, J. H. and E. Kreimer, *The Messier Album*. Cambridge, MA: Sky Publishing Corporation, 1978.

Mayer, Ben, *Starwatch*. New York: The Putnam Publishing Group, 1984.

Menzel, D. H. and J. M. Pasachoff, *A Field Guide to the Stars and Planets*, 2nd edition. Boston: Houghton Mifflin Co., 1983.

Mitton, S. (ed.) *The Cambridge Encyclopaedia of Astronomy*. New York: Crown, and London: Trewin Copplestone, 1977.

Moore, P. (ed.) *Guinness Book of Astro Facts and Feats*. London: Guinness, 1984.

Morrison, Philip and Phyllis, *Powers of Ten*. San Francisco: W. H. Freeman & Co., 1982.

Pasachoff, Jay M., *Astronomy from the Earth to the Universe*. Philadelphia: W. B. Saunders Co., 1979.

Vehrenberg, Hans, *Atlas of Deep Sky Splendors*. Cambridge, MA: Sky Publishing Corporation, 1978.

MAGAZINES

Astronomy/Deep Sky
Astromedia Corporation,
P.O. Box 92788, Milwaukee, W1 53202

Griffith Observer
Griffith Observatory,
2800 East Observatory Road, Los Angeles, CA 90027

Mercury
Astronomical Society of the Pacific,
1290 24th Avenue, San Francisco, CA 94122

Sky and Telescope
Sky Publishing Corporation,
49 Bay State Road, Cambridge, MA 02238

Telescope Making
Astromedia Corporation,
P.O. Box 92788, Milwaukee, WI 53202

STAR ATLASES

Norton's Star Atlas
Sky Publishing Corporation,
49 Bay State Road, Cambridge, MA 02238

Tirion Sky Atlas 2000.0
Sky Publishing Corporation and Cambridge University Press

AAVSO Variable Star Atlas
American Association of Variable Star Observers, 187 Concord Ave., Cambridge, MA 02138

TELESCOPES AND RELATED EQUIPMENT

Bushnell/Bausch and Lomb
2828 East Foothill Boulevard, Pasadena, CA 91107

Celestron International
2835 Columbia Street, Torrance, CA 90503

Coulter Optical
P.O. Box K, Idyllwild, CA 92349

Edmund Scientific
101 E. Gloucester Pike, Barrington, NJ 08007

Meade Instruments
1675 Toronto Way, Costa Mesa, CA 92626

Orion Telescopes
P.O. Box 1158-T, Santa Cruz, CA 95061

Questar Corporation
P.O. Box C, New Hope, PA 18938

Roger W. Tuthill
11 Tanglewood Lane, Mountainside, NJ 07092

STELAS AND STEBLICOM

Celestron International
2835 Columbia Street, Torrance, CA 90503

Index

AAT (Anglo-Australian Telescope), 117, 120
AAVSO, see American Association of Variable Star Observers
AAVSO Variable Star Atlas, 57, 106, 107
Abalone Galaxy (M64), 151
absolute magnitude, 18, 58–9
achromatic lens, 29, 31, 136
AF Geminorum, 57
Alcaino, Gonzalo, 166
Alcock, George E. D., 75, 103
Alcyone, 164
Algol, 73
alt-azimuth telescope mount, 15, 120
aluminum (mirror coating), 131, 133, 135
American Association of Variable Star Observers (AAVSO),
   information about, 55
   observations by, 58, 69
   photoelectric photometry, 147
   types of variable stars observed, 57, 72
   work of, 57, 107, 111
American Meteor Society, 107
American Meteoritical Society, 107
Andromeda, 162
Andromeda Galaxy (M31),
   distance, 17, 162
   like the Galaxy, 103, 105, 162; like M81, 155
   in our local group, 163
   novae, 103
   photos of, 138, 162
   size, 92, 162
   supernovae, 105
Andromeda Nebula, see Andromeda Galaxy
Anglo-Australian Telescope (AAT), 117, 120
anti-reflection coating, 135
Antares, 20
antennae, 27
antimatter, 77
aperture, 24, 36, 128, 133
aphelion, 17
Apollo 16, 84
apparent magnitude, 8, 17–18, 52, 57–8
Delta Aquarids, 63
Eta Aquarids, 63
Aquarius, 87, 88
Aquila, 58, 103
Arabic numbers (for star designation), 73
Araki, Genichi, 75
ARCTU (Automatic Recording Celestial Recording Unit), 143
Arcturus, 20
Arecibo Observatory (National Astronomy and Ionospheric Observatory), 117, 119

Aries, 8, 85–8
   first point of, 85, 87
Alpha Arietis (Hamal), 8
Aristotle, 11
Arizona, meteorite crater in, 77
Arizona, University of, 14
Arp, Halton C., 103
ASA, 35
Association of Lunar and Planetary Observers, 107
asteroids,
   'Ben Mayer' (No. 2863), 12, 60
   belt, 60, 77
   Ceres, 60, 65
   close approaches to Earth, 60, 74, 103
   discovery, 67, 102, 103; by blinking, 111
   Eros, 65
   Hermes, 60, 103
   Juno, 65
   occultation of a star, 60
   orbits, 12, 60, 77
   Pallas, 60, 65
   Vesta, 60, 65
astronomical unit, 17
atmospheric turbulence, 23, 30, 53, 92
Alpha Aurigids (Capellids), 63
Aurora Australis, 123
Aurora Borealis, 123
axis of the Earth, 47–8, 88–9, 93

B (camera setting), 25, 36, 47–8, 50, 141
B (magnitude), 20
Babylon, 8, 87, 88, 152
Bang-Bang Universe, 168–9
Barlow lens (tele-extender), 136, 149
Beckmann, Kenneth C., 103
Beehive (M44), 151, 154
Bell Laboratories, 15
Belt of Orion, 164
'Ben Mayer' (Asteroid No. 2863), 12, 60
Betelgeuse, 20
Big Bang Universe, 168–9
Big Dipper, see Ursa Major
binary system, eclipsing, 55, 57–8, 73
binoculars, 7, 25, 27
   comet hunting, 59, 65
   construction, 25
   magnitude limit, 17, 82
   objects for viewing, 113, 154, 158, 165
Blackeye Galaxy (M64), 151
blink comparator, 60, 108, 199, 111, 115
blinking, 50, 65, 94, 108–15
   discoveries, 60, 65, 103, 105, 114–15
   PROBLICOM, 81, 105, 110, 115

STEBLICOM, 81, 111, 112, 113
$B-V$, 20
boat trailer, 144
Bol'shoi Teleskop Azimutal'nya (Soviet 6-meter telescope), 15, 25, 31, 117, 120
Bond and Whipple, 41
Bonn (radiotelescope), 117
Bootids, 63
Zeta Bootids, 63
Bowell, Edward, 60
Bradfield, William A., 12, 112
Bryan, James, Jr., 105
Brigham Young University, 73
bulb (camera setting), 25, 36, 47–8, 50, 141

3C 324, 167
calculus, invention of, 11
Cambridge University, 12
camera, 22–33
   cold, 45, 136
   lenses, 23–5, 27, 79–80; dew on, 135, 139, 140. See also lenses
   35-millimeter, 80, 94, 111, 136; photography with, 7, 23, 27, 80, 114
   occulting shutter, 139, 141
   piggyback, 23, 127, 136
   pinhole, 23–5
   television, 25, 146–9. See also television
camera obscura, 22, 24
Canada – France – Hawaii Telescope (CFHT), 120
Cancer, 154
Canis Venatici, 157
carbon monoxide, 74
Carnegie, 120
Cassegrain, Louis, 31
Cassegrain telescope, 31–2, 122, 128, 133
Cassiopeia, 49
Beta Cassiopeids, 63
catadioptric systems, 32, 128, 130, 133, 140
CCD (charge-coupled device), 149, 166
CCTV (closed-circuit television), 146–9
celestial coordinates, 82–3, 85–7, 90–7. See also Equator, Celestial; Meridian, Celestial; Poles, Celestial
celestial mechanics, 12
celestial sphere, 82–3, 90–2
Celestron, 78, 139, 144
center of gravity, 11
Omega Centauri (globular cluster), 26, 27
Central Bureau for Astronomical Telegrams, 105–7

Alpha Cephei, 88
Delta Cephei, 72
Cepheid variable stars, 72, 73
Ceres, 60, 65
Cerro Tololo Interamerican Observatory (CTIO), 117, 120, 122–4
Cervit, 134
Omicron Ceti, (Mira), 57, 72
Chabot Science Center, 134
Chacaltaya Observatory, 117
Chambliss, Carlson R., 57
charge-coupled device (CCD), 149, 166
circumpolar stars, 85
cleaning optics, 135
Clippard solenoid, 142
clock drive, see motor driven mount
closed-circuit television, 146–9
closed Universe, 167–8
clusters,
   galactic (open), 136, 152; photos of, 75, 154, 159, 164
   of galaxies, 59, 105, 156
   globular, 26–7, 103, 136, 152; photos of, 26, 68, 158
   open (galactic), 136, 152; photos of, 75, 154, 159, 164
coated optics, 133, 135
cold camera, 45, 136
color balance (film), 37, 45
color index, 20
color sensitivity (film), 44
Comet Austin 1982, 98
Comet Bradfield 1979, 102
Comet Bradfield 1980, 12, 102
Comet Giacobini–Zinner, 12, 76
Comet Halley, 20, 59, 61, 74–6, 158
Comet Haneda–Campos 1978, 103
Comet Ikeya–Seki 1965, 60
Comet IRAS–Araki–Alcock 1983, 74–5, 103
Comet Kohler 1977, 99
Comet Swift–Tuttle, 76
comets, 59–60, 73–5
   collecting samples, 77
   coma, 73
   discovery, 51, 59–60, 67, 102, 109, 134; what to do with, 105–7
   'flying cocktail', 74
   Jupiter's family, 74, 76
   long-period, 12, 74, 99
   nucleus, 73
   as omens, 152
   orbits, 12, 74–5
   origin, 75
   periodic, 12, 74, 76, 99, 103
   photography, 59, 114, 136
   relationship to meteors, 61, 76
   seeker, 156
   short-period, 12, 74, 76, 103
   snowball, 75
   Sun-grazers, 60, 74, 102

tails, 73–4, 76, 98
theories, 74–5
wake, 61, 76
comparator, blink, 60, 108, 109, 111, 115
computers, 15, 72, 123, 125
constellations,
equatorial, 55
identifying, 49, 65, 93–4
meteor radiants in, 61, 63, 76
photography, 49, 55, 79, 82, 102, 111
polar, 49
starframes, 93–4
contracting Universe, 168
coordinate systems, celestial, 82–3, 85–7, 90–7
contrast (film), 44–5
converter, voltage, 145
Copernicus, Nicholai, 7,11, 15
Cornell University, 118
corona, solar, 53–5, 60, 74, 102
Corona Australis, 103
correcting lens (or plate), 32–3, 122, 127, 135
cosmic rays, 77, 117, 123
cosmology, 167
coudé focus, 123–4
Crab Nebula (M1), 72, 151, 152
craters,
on Earth, 77
on Moon, 53–4, 148
cross-hairs (reticle), 130, 135, 136
CTIO (Cerro Tololo Interamerican Observatory), 117, 120, 122–4
culminate, 92
Chi Cygni, 58
V 1500 Cygni, 73
XX Cygni, 73
Cygnids, 63
Cygnus,
Ben's target, 63, 67, 99
North American Nebula, 161
novae in, 72, 103, 111

Daguerre, Louis Jacques, 7, 41
daguerrotype, 7, 24
darkroom, 37–8, 41–2
declination,
definition, 83, 87, 91
guiding in, 130, 134
precession, 88
Delisle, J. N., 151
Deneb, 67
de Palma, Ralph, 35
developer, 41, 44
dew, 49, 135, 139, 140
dewguard, 139, 140
diaphragm,
camera, lens, 79
eye, 31
DIN, 35
distance–velocity law, 167–8
distance–red shift rule, 167–8
Dobsonian mount, 134
dolly lift, 144–5
Dog Star (Sirius), 17
Dombrowski, Phil, 60

double star, 72
Draconids, 76
Dreyer, John L. E., 152, 157
dual nature of light, 28
Dubhe, 92
Dumbbell Nebula (M27), 126, 151
du Pont telescope, 120
dust,
celestial, 26, 152, 165
terrestrial, 135
dwarf star, 69
Dwingeloo (Dutch radiotelescope), 117

Earth,
age, 169
atmosphere, 61
axis, 47–8, 85, 88–9, 93
Equator, 83, 85, 87–8, 91–3
latitude, 85, 87–8, 91–3
longitude, 83, 91
magnitude, absolute, 20
motion, 84–9
near approaches by asteroids, 60, 103
near approaches by comets, 74
nutation, 85, 88
orbit, 11, 12, 17, 72, 85–9
photo, 84
Poles, 47, 85, 88, 91–2
precession, 85, 88–9, 123
revolution (around the Sun), 47, 86–9
rotation (on its axis), 6, 37, 47–8, 79–81, 85, 90
–Sun relationship, 11, 73, 167
earthshine, 52
eccentricity, 11
eclipse, 152
solar, 53–5
lunar, 55, 56
eclipsing binary systems, 55, 57–8, 73
ecliptic, 87–8, 103
Einstein, Albert, 12, 168
Einstein's theory of relativity, 12–13, 169
electric drive, see motor driven mount
electronograph, 148
elliptical galaxies, 152, 162
elliptical orbits 11, 12, 17, 74
emulsion (photographic), 35, 40–1, 44–5, 53, See also film
Encyclopaedia Britannica Yearbook of science and the future, 64
enlarger, 37–8, 44, 127
epicycle, 11
Equator,
Earth, 83, 85, 87–8, 91–3
celestial, 83, 87, 91; photography near, 55, 79, 92
equatorial mount, 54, 88, 134, 136, 142, 144
equinox, 85, 87, 88
Eros, 65
ESO (European Southern Observatory), 120, 122, 166
Evans, Robert O., 103

expansion of the Universe, 15, 85, 167–9
exploding star, see novae; supernovae
external galaxies, see galaxies
eyepiece, 27, 29, 30–1, 128, 135–6

f (number, ratio, stop, or value), 24–5, 28, 53, 79–80
fall (autumn), 85–7
film, 35–7, 40–5
black-and-white, 35, 37–8, 41–5, 82
color, 35, 37–8, 44, 82
color balance, 37, 45
color sensitivity, 44
contrast, 44–5
developing, 37–8, 41–2
emulsions, 35, 40–1, 44–5, 53
grain (granularity), 40, 45, 53
hypersensitization, 41, 45
reciprocity failure, 36, 45, 61
speed, 24, 35–7, 44–5
fireball, 49, 50, 61, 67
first point of Aries (Vernal Equinox), 85, 87
fixer (hypo), 38, 41, 44
Flamsteed, John, 159
focal length, 25, 30, 32, 53, 134, 136
focal plane, 23, 27, 30, 32, 130, 147
focal ratio (or number, stop, value), 24–5, 28, 53, 79–80
focal scale, 29, 32, 53
focus, 25, 29–33
focussing, 139–40
forming gas, 41, 45
Frerichs, Denni, 134
fused quartz, 134

Gainsford, Eddie, 59
galactic (open) cluster, 136, 152
photos of, 75, 154, 159, 164
galactic nucleus, 85
galaxies, see also individual galaxies
atlas of, 105
cepheids in, 73
clusters of, 59, 105, 156
elliptical, 152, 162
Herschel and, 12
irregular, 152, 155, 157, 167
most distant measured, 167
nearest, 17, 20, 162, 163
novae in, 103
photographs of, 70, 155, 156, 157, 162, 163
photogarphy of, 127, 136, 151–3
recession of, 167–9
spiral, see under spiral galaxies
supernovae in, 58–9, 70, 103, 105
televised, 149
Galaxy (The Milky Way),
age, 169
appearance, 82, 152, 162, 169
cepheids in, 73
Earth's location, 15, 82
Earth's motion, 85, 88
galaxies nearby, 20, 163
globular clusters in, 26, 169

Herschel and, 12
novae in, 58, 99, 103, 105
photograph, 62
photography of, 41, 45, 82, 102, 111, 136
supernovae in, 58–9, 105
Galileo Galilei, 8, 11, 128, 154
gamma-rays, 15, 99
Garnavitch, Peter, 64, 69
gaseous nebulae, see nebulae
Gemini (constellation), 58
Geminids, 63
General Catalogue of Variable Stars, 106
globular clusters, 26–7, 103, 136, 152
photos of, 26, 68, 158
Gossamer radiotelescope, California 21
grain (granularity), 40, 45, 53
gravity, 11, 12, 167–9
Great Andromeda Galaxy, see Andromeda Galaxy
Great Bear, see Ursa Major
Great Nebula in Orion (M42), 44, 151, 153, 159
photo of, 165
Greenwich Observatory, 85, 91, 107
Greenwich Meridian, 83, 91
grinding (a mirror), 33, 128, 134–135
guider, automatic, 123, 136, 139
guiding, 123, 128, 130, 136, 141, 144
Gunter, J. U., 64

hairdrier, 135, 140
Hale, George Ellery, 120
Hale Observatory (Palomar; 200-inch telescope), 7, 13, 25, 27, 35, 117, 120
need for aperture, 25
appearance of globular cluster, 27
compared with other telescopes, 15
field of view, 32
focal length, 29
focal scale, 32
looking through, 7
photograph of, 13
vision of G. E. Hale, 120
Halley, Edmond, 158
Halley's comet, 20, 59, 61, 74–6, 158
Hamal, 8
Handbook of the Royal Astronomical Society, 106
Harvard University, 7, 41, 74, 115
hat trick, 49, 50, 141
heating pad, 125, 140
Hercules Cluster (M13), 151, 158
Hermes, 60, 103
Herschel, John, 15
Herschel, William, 12, 106, 115, 158
Hipparchus, 152, 154
hoist, 145
Honda, Minoru, 103, 106
Horseshoe Nebula (M17), 151
Hotel Cluny, 151
HST (Hubble Space Telescope), 16, 20
Hubble, Edwin, 15, 162
Hubble Space Telescope (HST), 16, 20
hydrogen, 159, 163, 167

hyperbola, 32
hypersensitization, 41, 45
hypo, 38, 41, 44

IAU Circulars, 59, 105–7
image, 27, 29, 31, 40, 148
image tube, 146–9
infrared radiation, 15, 99, 137
Infrared Astronomical Satellite (IRAS), 74–5, 103
Inquisition, 8, 11
instant camera, 38
intensifier tube, 146–9
International Astronomical Union (IAU), 59, 105, 107
International Ultraviolet Explorer (satellite), 105
interstellar dust, 26, 152, 165
interstellar matter, 103
IRAS (Infrared Astronomical Satellite), 74–5, 103
iris,
 camera lens, 36
 eye, 31
irregular galaxies, 152, 155, 157, 167
ISO, 35

Jansky, Karl, 15
Java, 53
Joner, Michael, 73
Juno, 65
Jupiter,
 beyond asteroid belt, 60, 77
 cloud belts, 54, 136, 148
 family of comets, 74, 76
 magnitude, 17
 orbit, 12
 satellites, 33

Kepler, Johannes, 11
Kepler's Star of 1604, 58
Kitt Peak National Observatory (KPNO), 57, 117
Kuwano, Yoshiyuki, 103

L³ TV (low-light-level television), 146, 148–9
Lagoon Nebula (M8), 151, 159
Large Magellanic Cloud (LMC), 20, 34, 103
Las Campanas Observatory, 103, 120
latitude (on Earth), 85, 87–8, 91–3
lawn-sprinkler timer, 63, 140–3
LED (light-emitting diode), 130
lenses,
 achromatic 29
 Barlow (tele-extender), 136, 149
 binocular, 27–8
 camera, 23–5, 27, 79–80; cost, 25; dew on, 135, 139–40; speed, 24–5, 79–80
 correcting (Schmidt), 32–33
 focal lengths, 7, 23, 25, 28–9, 111
 how they focus, 28–9
 magnitude limits, 82
 meniscus, 33
 refractors, 23, 27–9, 128

resolution, 30–1, 53, 55
scale, 29, 53
speed, 23–5, 35–6, 79–80, 127–8
telecompressor, 136, 149
tele-extender (Barlow), 136, 149
telelens, see telephoto lenses
telephoto, see telephoto lenses
Leo, 76
Leonids, 63, 76
Lick Observatory, 117, 120
light,
 dual nature, 28
 effect of gravity on, 168
 refraction, 28–9
 speed of, 8, 28, 167
light curve, 55, 57–9, 69, 73
light meter (photometer), 17, 122, 136, 146–7
light year, 8, 15
light-gathering power, 24
Lippershey (Dutch optician), 11
Little Bear, see Ursa Minor
Little Dipper, see Ursa Minor
LMC (Large Magellanic Cloud), 20, 34, 103
local group of galaxies, 163
log book, 35–8, 48–50, 114
longitude (on Earth), 83, 91
long-period comets, 12, 74, 99
Lowell Observatory, 60, 122
low-light-level TV (L³ TV), 146, 148–9
lunar, see Moon
Lyra, 73, 160
3 Lyrae (Vega), 73
Alpha Lyrae (Vega), 17, 20, 73, 88
Beta Lyrae, 160
Gamma Lyrae, 160
Lyrids, 63

M1 (Crab Nebula), 72, 151, 152
M8 (Lagoon Nebula), 151, 159
M13 (Hercules Cluster), 151, 158
M17 (Omega or Horseshoe Nebula), 151
M20 (Trifid Nebula), 116, 151
M24, 39
M27 (Dumbbell Nebula), 126, 151
M31 (Andromeda Galaxy), 138, 151, 152, 162. See also under Andromeda Galaxy
M32, 162
M33 (Triangulum Spiral), 163
M42 (Orion Nebula), 44, 78, 151, 153, 159, 165
M44 (Praesepe or Beehive), 75, 151, 154
M45 (Pleiades), 37, 38, 151, 164
M51 (Whirlpool Galaxy), 151, 157
M57 (Ring Nebula), 151, 160
M64 (Blackeye or Abalone Galaxy), 151
M69, 151
M81, 155
M82, 151, 155, 167
M87, 105
M92, 68
M97 (Owl Nebula), 151, 152

M100, 70
M104 (Sombrero Galaxy), 129, 151, 156
Machholz, Donald E., 102
Magellanic Clouds, 20, 103
magnification, 23–4, 27–8, 30–1
magnitude, 17, 56–8, 82
 absolute, 18, 58–9
 apparent, 8, 17–18, 52, 57–8
Maksutov, D. D., 32
Maksutov telescope, 32–3
Manchester (English radiotelescope), 117
Mars,
 inside asteroid belt, 60, 77
 comets beyond, 73
 magnitude, 17
 orbit, 12, 73, 86
 retrograde motion, 86, 113
Marsden, Brian G., 105
Mauna Kea (Hawaii), 117, 118, 120
Mayan priests, 88
Mayer, Lucille, 125
Meade, 139
Mechain, Pierre, 156
Meier, Rolf, 102
meniscus lens, 33
Merak, 92
Mercury, 52
 comets inside orbit of, 74
 orbit, 11, 12
Meridian, Celestial,
 definition, 83, 85, 87
 photography near, 142
 sidereal time, 87, 152–3
Meridian Greenwich, 83, 91
Messier, Charles, 103, 151–2, 156–8, 160–1
Messier Catalogue, 151–2, 156
meteorites, 61, 77
meteors, 61, 75–6
 altitude, 61, 75–6
 origin, 76
 photography, 37, 49–51, 61, 63, 80, 143, 150
 showers, 61, 63, 65, 76–7; table of, 63
 size, 61, 76
 sporadic (random), 63, 76
 velocity, 76, 150
Michigan University of, 122
Mikesic, Dragan, 133
milk glass, 64
Milky Way, see Galaxy
minor planets, see asteroids
Mira, 57, 72
mirror (telescope), 23, 31–3, 127–8
 care of, 135
 compared to lens, 27, 128
 Herschel's, 12
 home-made, 33, 128, 134–5
 Messier's, 151
 metallic, 32, 151
 Newton's, 12
 shape of, 12, 31–33
MMT (Multiple Mirror Telescope), 14, 15, 120
monocular, 27
Moon,
 craters, 53, 55, 148

eclipse, 55, 56
magnitude, 17
mountains, 55
occultations by, 55, 67
orbit, 88
photography, 25, 36, 52–3, 55, 113, 136
size, angular, 92; linear, 53
televised, 148–9
X-rays from, 15
motor-driven mount, 80–2, 88, 134
 ARCTU, 143–4
 guiding with, 129–30
 photography, 93, 111, 127, 134, 142
 setting on stars, 93
STELAS, 80–1
Mount Everest, 91
Mount Hamilton (Lick Observatory), 117, 120
Mount Hopkins (MMT, Whipple Observatory), 14, 15, 120, 139
Mount Palomar, see Palomar Observatory
Mount Pastukhov (Soviet 6-meter telescope), 15, 25, 31, 117, 120
Mount Wilson Observatory (100-inch telescope), 15, 103, 117
mouse-trap (shutter), 105
Multiple Mirror Telescope (MMT), 14, 15, 120
Mu Muscae, 105

NASA (National Aeronautics and Space Administration), 117
National Astronomy and Ionospheric Observatory (Arecibo Observatory), 117, 118
National Radio Astronomy Observatory (NRAO), 17, 117
National Science Foundation (NSF), 117, 123
NCP (North Celestial Pole), see Poles, Celestial
nebulae, see also individual examples
 appearing like comets, 59, 151–2
 around novae, 72
 gaseous, 12, 72, 136
 Herschel and, 12
 photographs, 158–61, 164, 165
 photography of, 136
 planetary, 151, 152, 160
 televised, 149
Neptune, 12, 113
neutron star, 72
New General Catalogue (NGC), 152
new stars, see novae
Newton, Isaac, 8, 12, 31
Newtonian reflector, 31–2, 130
 best buy, 128, 133
 Herschel's, 12
 home-made, 169
 Messier's, 151
NGC (New General Catalogue), 152
NGC 205, 162
NGC 224 (M31), 151, 152, 162. See also Andromeda Galaxy
NGC 253, 10

NGC 598 (M33), 163
NGC 1952 (M1), 72, 151, 152
NGC 1976 (M42), 44, 151, 153, 159, 165
NGC 2632 (M44), 75, 151, 154
NGC 3031 (M81), 155
NGC 3034 (M82), 151, 155, 167
NGC 3587 (M97), 151, 152
NGC 4594 (M104), 151, 156
NGC 5194 (M51), 151, 157
NGC 6205 (M13), 151, 158
NGC 6523 (M8), 151, 159
NGC 6720 (M57), 151, 160
NGC 7000 (North American Nebula), 161
Niedner, Malcolm B., Jr, 75
non-periodic comets (long-period comets), 12, 74, 99
North American Nebula, 161
North Celestial Pole (NCP), see Poles, Celestial
Northern Cross, 49, 63, 99
North Pole (terrestrial), 47, 85, 88, 91-2
North Star, see Polaris
Norton's Star Atlas and Reference Handbook, 73, 82, 114
notebook, 35-8, 48-50, 114
Nova Aquila 1982, 103
Nova Cygni 1975, 62-7
  discovery, 58, 63
  theory, 69, 72
  photograph, 67
  photography, 51, 63-7
Nova Cygni 1978, 103
Nova Muscae 1983, 59, 104, 105
Nova Normae 1983, 107
Nova Trianguli 1983, 107
novae,
  color, 69
  description, 57-8
  discovery, 49, 55, 63-5, 105, 107
  fast, 59, 64
  in other galaxies, 103
  light curves, 59, 69
  in Milky Way, 58, 103, 111
  photographs, 67, 104
  photography, 49-51, 63-7
  recurrent, 72, 103
  shells around, 72
  size, 72
  slow, 59
  temperature, 69
  theory, 69, 72
NRAO (National Radio Astronomy Observatory), 17, 117
NSF (National Science Foundation), 117, 123
nuclear reaction, 69
nucleus,
  atomic, 167
  of the Galaxy, 55
nutation, 85, 88

objective, (lens), 27-30
Observer's Handbook of the Royal Astronomical Society of Canada, 106

occultations of stars
  by asteroids, 60
  by the Moon, 55, 67
occulting shutter, 109-10, 139, 141
ocular (eyepiece), 27, 29, 30-1, 128, 135-6
off-axis guiding, 130
Omar Khayyam, 82
Omega Nebula, 151
open (galactic) cluster, 136, 152
  photos of, 75, 154, 159, 164
open Universe, 168
optics, 12, 26-33
orbits,
  asteroids, 12, 60, 77
  comets, 12, 74-5
  Earth, 11, 12, 17, 72, 85-9
  planets, 11-12, 60, 73, 77, 86
19 Ori (Rigel), 73
Orion,
  belt, 165
  location, 87, 153
  Nebula (M42), 44, 151, 153, 159; photos, 78, 165
  photography, 49, 55
  star names in , 73
  Trapezium, 165
  viewing, 93
Orionids, 61, 63, 76
Beta Orionis (Rigel), 73
Osada, Kentaro, 64, 65, 99
Overbeek, M. D., 105
overcoat (mirror), 133, 135
Owl Nebula (M97), 151, 152

Pallas, 60, 65
Palomar Observatory, 7, 117
Palomar Schmidt telescope, 120
panchromatic emulsion, 44
Panther, Roy W., 102
parabola, 12, 31-2, 117, 135
Parkes (Australian National Radio Astronomy Observatory), 117
parsec, 15
Payne-Gaposchkin, Cecilia, 103
perihelion, 17
periodic comets, 12, 74, 76, 99, 103
period-absolute magnitude relation, 73
Beta Persei (Algol), 73
Perseids, 63, 76
Peterson Ruth, 107
Phobos, 148
phonograph record, 141
photocell (phototube), 136, 141, 147
photoelectric,
  photometer, 17, 122, 136, 146-7
  switch, 140
photographic magnitude, 20
photography, *see* camera; film; lenses
photometer, 17, 122, 136, 146-7
photomultiplier (PM), 147
photon, 23, 28-30, 128, 149, 168
phototube (photocell), 136, 141, 147
physics, laws of, 168
Pic du Midi Observatory, 117
picture element (pixel), 149
pinhole camera, 23-5

Pisces, 55, 87, 88
Lambda Piscium, 87
Omega Piscium, 87
pixel (picture element), 149
PKS 2000-330, 166, 167, 169
planetary nebula, 151, 152, 160
planets, *see also* individual examples
  motions, 11-12, 86, 113-14
  orbits, 11-12, 17, 60, 72, 73, 77, 85-9
  unlike planetary nebulae, 152, 160
plates, photographic, 44, 45, 111
Pleiades (M45), 37, 38, 151, 164
Plough *see* Ursa Major
Pluto, 11-12, 74, 75
PM (photomultiplier), 147
pointer stars, 47-8, 92
polar axis, 80-2
  alignment of, 81-2, 88, 93, 127, 130, 153, Appendix
Polaris,
  Earth viewed from, 91
  location, 47-8, 85, 92
  magnitude, 20
  motion around, 46, 48, 55
  motion of, 85
  photo of, 46
  photography of, 46-50, 79, 92-3
  in polar alignment, 81, Appendix
  as pole star, 88-9
Poles, Celestial,
  defined, 47
  location of, 85, 87
  photography of, 48, 55, 61, 79, 92-3
  polar alignment, *see under* polar axis
  view from, 83, 91
Poles, terrestrial, 47, 85, 88, 91-2
Pole Stars, *see* Polaris; Sigma Octantis
pollution, light, 36, 82
Pollux, 20, 58-9
polyethylene bag, 49, 140, 142, 144
Praesepe (M44), 75, 151, 154
precession, 85, 88-9, 123
primary mirror, 31-3, 128
prime focus, 123
prism, 8, 12, 35
PROBLICOM (Projection Blink Comparator), 81, 105, 110, 115
proton, 167
Proxima Centauri, 15
Ptolomy, 113
pulsar, 15
pyrex, 134

Quadrantids, 63
quasar, 15, 99, 166-9

R.A., *see* Right Ascension
radiant, 61, 76
radiation pressure, 74
radioactivity (in meteorites), 61, 77
radioastronomy, 15, 99, 117
radiotelescopes, 15, 17, 117-18
random (sporadic) meteors, 63, 76
reciprocity failure, 36, 45, 61
recurrent novae, 72, 103
Red Planet (Mars), 113

red shift, 167-8
reflecting telescopes, 12, 31-2; *see also* Cassegrain telescope; Maksutov telescope; Schmidt telescope; Schmidt-Cassegrain telescope
  care of, 135
  dew guard, 140
  focussing device, 139
  Herschel's, 12
  light-gathering power, 24
  making, 33, 128, 134-5
  Messier's, 151
  Newton, 12
  photography with, 136
  relative prices, 133
  for terrestrial viewing, 23
refracting telescopes, 27-30, 128, 133, 140, 147
refraction, 27-30
relativity, theory of, 12, 13, 169
resistors, 140
resolution, 23, 31, 54
resolving power, 30-1, 53
reticle (cross-hairs), 130, 135, 136
retrograde motion, 86, 113
Rigel, 73
Right Ascension (R. A.),
  definition, 83, 85, 91
  guiding in, 130, 134
  how to determine, 152-3
  using in photography, 111, 114, 142
  precession, 88
  sidereal time, 87, 152
  starframes, 94
Ring Nebula (M57), 151, 160
Roman letters (star designation), 73
Royal Astronomical Society, 106, 152
Royal Greenwich Observatory, 85, 91, 107

safe-light, 38, 41, 44
Sagitta, 103
WZ Sagittae, 103
V 3857 Sagittarii, 103
Sagittarius,
  location, 93
  Messier objects in, 151, 159
  novae in, 58, 103, 111
satellites,
  astronomy from, 15, 65, 99, 149
  Jupiter's, 33
  photography of, 41, 56
  view of Earth from, 91
Saturn, 12, 148, 156
  rings of, 33, 54, 136, 148, 156
scale, focal, 29, 32, 53
SCAMP (Self-Contained Astronomical Mobile Platform), 143-5
Schliemann, Heinrich, 9
Schmidt, Bernard, 32
Schmidt camera (or telescope), 32-3
  at CTIO, 122
  at ESO, 120
  M33 photographed with, 163
  at Palomar Observatory, 120
  of the University of Michigan, 122
  versus telelens, 127-9

Schmidt–Cassegrain telescope, 32–3, 169
Schucking, E. L., 167
scissor jack, 136
Scorpius, 58, 87, 93
Sculptor, 10
seasons, 87–8
seconds of arc, 15
secondary mirror, 31–3, 122, 125, 128
seeing, 53–4
Serpens, 103
seven sisters (Pleiades), 164
Shapley, Harlow, 15
shell, nova, 72
shooting stars, see meteors
sidereal time, 87, 152–3
Siding Spring (Anglo-Australian Telescope), 120
Sigma Octantis, 89. See also Poles, Celestial
silicon vidicon (L³ TV), 146, 148–9
silver,
  in film, 41
  mirror coating, 135
silver bromide (or chloride), 41
Sirius, 17
slide projectors, 27, 109–11
Small Magellanic Cloud (SMC), 20
Smithsonian Institution (Astrophysical Observatory), 14, 61, 63–4, 74, 106–7, 120
solar, see Sun
Solar System,
  like a giant atom, 167
  center of gravity, 11
  comets in the, 59, 73–4
  Copernicus, 7, 11
  diagrams, 12, 74, 86, 89
  in the Galaxy, 82
  Mira's diameter compared with, 72
  plane of, 88
  space astronomy and, 15
soldering iron, 140
solenoids, 63, 141–3
solstice, 87–8
Sombrero Galaxy (M104), 151, 156
sound, speed of, 8, 72
South Celestial Pole (SCP), see Poles, Celestial
Southern Cross, 17, 105
South Pole (terrestrial), 47, 85, 88, 91
Soviet 6-meter telescope, 15, 25, 31, 117, 120
Space Telescope (Hubble Space Telescope), 16, 20
space telescopes, 15, 16, 20, 149
spectrograph, 74, 105, 122–3, 137
spectroscope, 137
spectroscopy, 8, 136, 137
spectrum (plural spectra), 35, 107, 137
speculum, 151
speed (of a lens), 23–5, 35–6, 79–80, 127–8
speed of light, 8, 28, 167
spherical mirror, 32–3

spiral galaxies, see also individual examples
  distance to M31, 17, 162
  LMC, 20
  in Messier's catalogue, 152
  novae in, 103
  photographs, 10, 20, 70, 131, 155–7, 162–3
  photography of, 127
  supernovae in, 70
Spiral Galaxy in Triangulum (M33), 163
sporadic (random) meteors, 63, 76
spring (season), 85–7
SS Cygni, 72
STARFRAMES™, 93–7
stars,
  binary, 57, 58, 73
  clusters, 127, 136, 151. See also clusters
  color index, 20
  colors, 20, 69, 72, 82
  diameters, 72
  distances to, 8, 15, 17, 20, 73
  double, 72
  dwarf, 69
  eclipsing, 55, 57–8, 73
  giant, 72
  nearest, 15, 17
  spectra, 35, 107, 137
  temperature, 69, 72
  tracker, 123, 136, 139
  trails, 37, 47–51, 55, 60; magnitude limit, 58; photos of, 7, 46, 56
  variable, 57–9, 73, 102, 111. See also variable stars
Steady State Universe, 168
STEBLICOM (STEreo BLInk COMparator), 81, 111–13
STELAS (Stop The Earth, Lock All Stars), 80–1
STELASCOPE, 80–1
stepping motor, 134
stereopsis, 111
summer, 88, 124
Sun,
  absolute magnitude, 20
  age, 8
  apparent magnitude, 17
  color index, 20
  corona, 53–5, 60, 74, 102
  dangers of photographing, 36, 53
  distance, 11, 17
  –Earth system, 11
  eclipse, 8, 53–5
  location in the Galaxy, 15, 82
  motion in space, 85, 88
  path in sky, 85–8
  prominences, 55
  seasons, 87–8
  in Solar System, 11–12
  sungrazing comets, 60, 74, 102
  temperature, 69, 72, 137
  time, 87
  wind, 74
sungrazing comets, 60, 74, 102
supernovae, 57–9, 72

Crab Nebula, 72, 151, 152
  discovering, 50, 57, 103–4, 107, 115
  frequency, 58–9, 105
  photograph, 70
  theory, 72
  where found, 58–9

Taurus, 22, 37, 86, 88, 164
telecompressor, 136, 149
tele-extender (Barlow lens), 136, 149
telelens, see telephoto lenses
telephoto lenses,
  cost, 25
  field of view, 111
  focal lengths, 25
  focal scale, 29
  lunar photography with, 53–4
  magnitude limit, 82
  piggyback, 23, 127, 136
  versus Schmidt, 127–8, 130
telescope making, 33, 128, 133–5, 169
telescopes, 15, 26–33, 132–7. See also individual types
television, 146–9
  cameras, 25, 146
  closed-circuit (CCTV), 146, 147–9
  with image tube, 149
  low-light-level (L³ TV), 146, 148–9
  silicon vidicon, 148
  for showing one's photos, 114
  on telescope, 123, 136, 148
temperature (Sun, stars), 69, 72, 137
Thompson, Gregg, 105
Thuban, 88
time,
  exposure, 35–7, 48–9, 53
  sidereal, 87, 152–3
  solar, 87
time machine, 169
timer, 63, 140–3
Tirion Sky Atlas 2000.0, 73, 82, 93
trailer, boat, 144
transformer, 140–2
Trapezium, 165
Trifid Nebula (M20), 116, 151
Tunguska, 77
TV, see television

UBK-7, 134
ultraviolet radiation, 15, 137
Universe, 166–9
  age, 169
  Bang-Bang, 168–9
  Big Bang, 168–9
  closed, 167–8
  expansion, 15, 85, 167–9
  open, 168
  origin, 168–9
  Steady State, 168
Upton, Edward, K., 64
Uranus, 12, 15, 111, 113, 115
Ursa Major,
  brightness of stars, 17
  description of, 47–8
  M81 and M82, 155
  motion of, 48, 85

other names, 48
photograph of, 6
photography of, 49
pointer stars, 47–8, 92
Ursa Minor, 46–8, 85, 93

V (magnitude), 20
vacuum, 28
variable stars, 57–9
  blinking for, 111
  cataclysmic, 57
  cepheids, 72–3
  discovery 103, 106
  eclipsing binary stars, 55, 57–8, 73
  exploding, 57, 59. See also novae; supernovae
  light curves, 57, 58, 59, 69, 73
  long-period, 57, 72
  Mira, 57, 72
  nomenclature, 73
  novae, see novae
  photography, 55
  pulsating, 57, 59
  short period, 55
  study of, 102
  supernovae, see supernovae
Vega, 17, 20, 73, 88
Venus, 12, 17, 54, 74, 148
Vernal Equinox (first point of Aries), 85, 88
Very Large Array (VLA), 17
Vesta, 60, 65
vidicon, silicon (L³ TV), 146, 148–9
Virgil, 88
Lambda Virginis, 115
Virgo, 55, 93, 115, 131, 156
Virgo Cluster, 105, 156
visual magnitude, 17–18, 69
VLA (Very Large Array), 17
Vulpecula, 103

Walden Pond, 52
wavelength, 15, 133, 134
wheatgrain, 130
Whipple (of Bond and Whipple), 47
Whipple, Fred L., 74
Whipple (Fred L.) Observatory, 103, 139
Whirlpool Galaxy (M51), 151, 157
winter, 88, 125

X-rays, 15

Yale University, 122
year, 87–88
Yearbook of Science and the Future (Encyclopaedia Britannica), 64
Yeomans, Donald K., 75
Yerkes Observatory, 31, 120

Zelenchuskaya (Soviet 6-meter telescope site), 25, 117, 120
zenith, 74, 87–8, 92
zodiac (ecliptic), 60, 87–8, 103
zoom eyepiece, 30, 135